# Nonimaging Optics in Solar Energy

Nonimaging Optics in Solar Energy

Joseph J. O'Gallagher

ISBN: 978-3-031-79419-3    paperback

ISBN: 978-3-031-79420-9    ebook

DOI: 10.1007/978-3-031-79420-9

A Publication in the Springer series

*SYNTHESIS LECTURES ON ENERGY AND THE ENVIRONMENT:*

*TECHNOLOGY, SCIENCE, AND SOCIETY*

Lecture #2

Series Editor: Frank Kreith, Professor Emeritus, University of Colorado

**Series ISSN**

ISSN 1940-851X        print

ISSN 1942-4361        electronic

# Nonimaging Optics in Solar Energy

**Joseph J. O'Gallagher**
Alternative Energy Solutions
Flossmoor, Illinois

*SYNTHESIS LECTURES ON ENERGY AND THE ENVIRONMENT: TECHNOLOGY, SCIENCE, AND SOCIETY #2*

# ABSTRACT

Nonimaging optics is a subdiscipline of optics whose development over the last 35–40 years was led by scientists from the University of Chicago and other cooperating individuals and institutions. The approach provides a formalism that allows the design of optical devices that approach the maximum physically attainable geometric concentration for a given set of optical tolerances. This means that it has the potential to revolutionize the design of solar concentrators. In this monograph, the basic practical applications of the techniques of nonimaging optics to solar energy collection and concentration are developed and explained. The formalism for designing a wide variety of concentrator types, such as the compound parabolic concentrator and its many embodiments and variations, is presented. Both advantages and limitations of the approach are reviewed. Practical and economic aspects of concentrator design for both thermal and photovoltaic applications are discussed as well. The whole range of concentrator applications from simple low-concentration nontracking designs to ultrahigh-concentration multistage configurations is covered.

# KEYWORDS

solar thermal energy, photovoltaic conversion, nonimaging optics, optical concentration, high temperature, thermal energy conversion, secondary concentrator, compound parabolic concentrator, "trumpet concentrator"

# Acknowledgments

Much of the research whose results are described here was supported in part by the U.S. Department of Energy through the Office of Basic Energy Sciences, the Office of Energy Efficiency and Renewable Energy, Sandia National Laboratory and the National Renewable Energy Laboratory under a variety of grants and contracts. I express particular thanks to my good friend and colleague Roland Winston, who invented the field of nonimaging optics and provided me with the opportunity to experience the exciting period of its early applications in Solar Energy. He continues to lead its development now at the University of California, Merced. Others deserving special mention include Peretz Greenman, Ari Rabl, Manuel Collares-Pereira, Philip Gleckman, Dave Jenkins, Keith Snail, Xiao Ning, the late David Cooke, Deivarajan Suresh, Dan Sagie, Eli Kritchman, Walter Welford, Harald Reis, and Julius Mushaweck all of whom spent time as students or visiting researchers at the University of Chicago; I convey my appreciation to Bill Schertz and John Hull of the Argonne National Laboratory; Al Lewandowski and Carl Bingham of the NREL Rich Diver and Rod Mahoney at Sandia National Laboratory; and Bill Duff of the Colorado State University. Finally, I must thank all the other, too-numerous-to-mention but certainly special individuals, support staff, and students who have worked with me over the past 30 years and more to develop and apply the concepts of nonimaging optics to solar energy concentration.

# Important Note

Many of the concepts described herein are covered by one or more patents issued to Roland Winston and others who worked with him at the University of Chicago. The rights to the body of intellectual property covered by these patents are held for the most part by one of several companies that are subsidiaries of Solargenix Energy, LLC. Individuals or organizations interested in the commercial development and application of associated devices should contact Brian Voice of Raleigh, North Carolina.

# Contents

CHAPTER 1

# Introduction

## 1.1 WHAT IS "NONIMAGING OPTICS?"

Nonimaging optics is a new approach to the collection, concentration, and transport of light developed by physicists from the University of Chicago over the past 35 years. See, for instance, the work of O'Gallagher and Winston (1983), Welford and Winston (1978; 1989), and Winston (1974; 1991; 1995, Ed.). The basic idea is to relax the constraints of point-to-point mapping of imaging optics, which are not essential if one's goal is to collect as much light as possible. This in turn permits the design of optical systems that achieve or closely approach the maximum geometric concentration permitted by physical conservation laws for a given angular field of view. Not surprisingly, this has important consequences for the design of solar concentrators. Of particular note is the fact that, prior to the development of this approach, it was accepted conventional wisdom in the field of solar concentrator design that no useful concentration could be achieved without tracking the collector (moving it to follow the sun). This has been shown to be untrue. Application of the techniques of nonimaging optics provides designs that deliver moderate levels of concentration with completely stationary concentrators. Higher concentrations do require tracking, but, at any level, use of nonimaging techniques allows greatly relaxed optical tolerances and/or the achievement of still higher concentrations with the same tolerances.

The compound parabolic concentrator (or CPC as it is often called; Rabl, 1976b) is the prototypical "ideal" nonimaging light collector. It was invented by Roland Winston (1974), and the term "CPC" has become generic for the whole family of similar devices. (It is interesting to note that the retinal receptors in the eyes of the horseshoe crab, a creature that evolved several hundred million years ago, have been found to have the shape of CPCs (Levi-Setti, Park, and Winston, 1975).)

There are several important differences between nonimaging optics and conventional "imaging" or focusing optics. The underlying formalism of nonimaging optics analyzes light propagation in terms of phase space quantities and/or energy flow patterns. In nonimaging optics, the optical surfaces (mirrors, lenses) are designed for the extreme angular rays of the desired field of view rather than for axial rays. Rays closer to the axis are not brought to a focus, but all are still collected. These nonimaging systems can approach and, in some cases, attain the so-called thermodynamic limit for concentration. Note that focusing systems always fall short of this limit (by a factor of 3–10).

## 1.2    ADVANTAGES FOR SOLAR ENERGY CONCENTRATION

Nonimaging optics achieves the widest possible angular field of view for a given geometric concentration. As noted earlier, this permits useful concentration without tracking. Low to moderate concentrations (1.1×–2×; in special cases, up to 4×) can be achieved with a totally stationary (fixed year-round) collector. Slightly higher levels (~ 3×–10×) will usually require occasional (seasonal) adjustment. Another advantage is that such systems collect a substantial fraction of the "diffuse" component of insolation, much of which is lost in conventional focusing concentrators.

Higher concentrations (>10×–40,000×) will require tracking. However, with nonimaging techniques, these levels can be attained with relaxed optics and tracking requirements. These generally lead to simpler, less expensive, and more easily maintained concentrator systems. This is particularly important for developing economical and cost-effective solar collector systems. Furthermore, although the diffuse component of insolation cannot be collected at these concentration levels, much more of the near-sun "circumsolar" radiation will be concentrated with this approach.

In general, wherever concentration of sunlight is desired, nonimaging optics can achieve the highest possible levels with the most relaxed optical tolerances. These in turn have great potential for attaining economic viability for solar energy conversion in a wide variety of applications, from rural applications in developing countries to very high-technology applications, such as hydrogen production and laser pumping in space.

## 1.3    DEFINITION OF GEOMETRIC CONCENTRATION FACTOR

Before proceeding any further, we need to define what we mean by "geometric concentration factor." A solar concentrator is an optical device that collects the solar radiation incident on some aperture area $A_1$ and delivers it to a (usually smaller) absorber area $A_2$. The geometric concentration ratio is then defined to be

$$C_{geom} = A_1 / A_2 \qquad (1.1)$$

### 1.3.1    Concentration and the Thermodynamic Limit

It is easily shown that, for the definitions illustrated in Figure 1.1, the maximum geometric concentration allowed by physical conservation laws is related to the angular field of view by

$$C_{max} = \frac{n}{\sin\theta} \qquad (1.2a)$$

in two dimensions (trough-like, or translational geometry perpendicular to the page) and

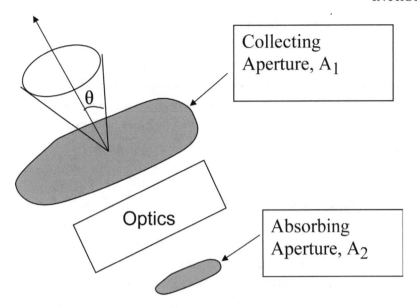

**FIGURE 1.1:** Definition of geometric concentration. For $C_{geom} > 1$ (i.e., for $A_2 < A_1$), the optics must limit the field of view, defined for simplicity here by a cone of half-angle $\theta$.

$$C_{max} = \frac{n^2}{\sin^2\theta} \qquad\qquad (1.2b)$$

in three dimensions (cone-like, or rotational geometry around the optical axis).

Here, $n$ is the index of refraction at absorber surface and $\theta$ is the half-angle of acceptance.

By convention, any system that can attain these limits is referred to as "ideal." By "attain these limits" we mean that the whole light incident on the collecting aperture $A_1$ within the acceptance half-angle $\theta$ is delivered *without geometric losses* to the absorber $A_2$. As a corollary to this behavior, this means that *none* of the light incident outside the acceptance half-angle $\theta$ makes it to the absorber (Collares-Pereira, O'Gallagher, and Rabl, 1978). Please note that all conventional imaging systems fall short of this limit by a factor of at least 2–4 (Rabl, 1976a).

There are two usual proofs of this limit. The first is based on a thermodynamic argument and shows that if $C_{geom}$ could be made larger (absorber $A_2$, smaller), there would not be enough area to radiate away the incident energy and its temperature could rise above that of the source in violation of the second law. The second proof is based on the conservation of the phase space volume occupied by a bundle of light rays. This is analogous to the Liouville Theorem for light, namely, that "brightness" is conserved along a light ray. For more rigorous treatment of these concepts, see, for example, the work of Rabl (1976a) or Welford and Winston (1978).

Of course there are further practical limits to the *net* concentration that can be achieved with a real optical device due to reflection and transmission losses in the less-than-perfect material properties used in the device. We shall be concerned with these kinds of inefficiencies, which we will want to minimize, when we are designing real solar concentrators. However, it must be recognized that the limits of Eq. (1.2) are fundamental and apply even in the idealized limit of perfectly reflecting and perfectly transmitting optical materials. The losses in this case would be geometric losses in throughput.

## 1.4   THE ROLE OF CONCENTRATION IN SOLAR ENERGY CONVERSION

The definition of the geometric concentration ratio (Eq. (1)) is quite simple and straightforward, yet, surprisingly, it is very often misunderstood. It should be emphasized that when used for solar energy conversion, $C_{geom}$ is often NOT a linear figure of merit: That is, a geometric concentration of, say, 10× is not necessarily two times better than a geometric concentration of 5×. It depends on the purpose for which the concentration is employed. To understand this better, we need to consider what the only two motivations for using concentration in solar collectors really are. These are (1) to improve performance and (2) to improve the economics by reducing cost. Each of these purposes has a different relative importance for the two major solar energy conversion technologies—that is, solar thermal and solar photovoltaic (PV) applications.

The performance motivation can be best understood in the context of the solar thermal application. Here, the main purpose of concentration is to increase the efficiency (i.e., the performance) at higher temperatures by reducing the heat losses. In any solar thermal collector, the heat losses are directly related to the area of the hot absorber—the smaller this area, the smaller the heat losses. In fact, for flat plate collectors, the heat losses at operating temperatures much above those of boiling water (100 °C) are so large that very little or no useful heat can be extracted for the desired end use. The efficiency is very low or even approaches zero. Introducing concentration, so that the area of the hot absorber is only a fraction of the aperture, reduces the thermal losses at a given operating temperature, roughly by a factor of $1/C_{geom}$ on an aperture area basis. In this way, respectable operating efficiencies can be attained at higher and higher temperatures. This is the reason that parabolic troughs or dishes are often employed to generate the temperatures required to drive heat engines. We shall see that nonimaging optics, combined with other heat loss reduction techniques, such as spectrally selective surfaces and vacuum insulation, can also provide good performance at such temperatures.

On the other hand, the economic motivation for employing concentration can be best understood in the context of PV applications. Here, the major purpose of employing concentration is to reduce overall system cost. Solar cells are quite expensive, and if one has to cover large areas with these expensive transducers, the cost can be prohibitive. However, if one can reduce the area of expensive material per unit aperture area, this cost can be substantially reduced. In the limit that the optics would be free, the cost could be reduced by a factor of $1/C_{geom}$. Of course the optics is not free. However, if inexpensive reflective mirrors or lenses can be used, they can be much less expensive per unit aperture than the expensive cell material. In any event, at higher and higher concentrations, the relative cost of the energy transducer per unit aperture becomes small and attention can then be directed toward developing inexpensive concentrators rather than inexpensive cells.

The performance motivation for concentration in PV systems is less important. In fact, use of concentration often introduces factors (e.g., higher cell operating temperatures and nonuniform illumination of the cells) that threaten to degrade performance. If such effects can be overcome, the efficiency of PV conversion of sunlight to electricity does increase slowly (approximately logarithmically) with increasing flux. This potential for increased performance does serve to offset some of the complicating effects of concentration mentioned earlier and does motivate striving for even higher levels of concentration in some systems under development. However, the primary motivation for concentrating PVs is to reduce system cost.

Finally, we note that the competing economic impacts of employing concentration in solar thermal systems are not even often recognized. Although the driving purpose for concentration is to attain high temperatures (by reducing heat losses), the economic effects become increasingly notable at very high concentrations when the costs of the thermal receivers that are required at the resulting high operating temperatures begin to be appreciable. There is a type of reverse synergism operating here. Concentration is employed to reach high operating temperatures. This in turn drives the receiver cost upward to the point in which it is necessary to collect the solar energy from a very large area to make it worth that cost. It becomes very difficult to design a cost-optimized system under such conditions, and this may explain in part why high-concentration solar thermal systems (e.g., central receivers and dish–Stirling systems) are still struggling to approach economic viability. We shall return to these issues later.

C H A P T E R   2

# CPCs

## 2.1   BASIC GEOMETRY

The geometry of a simple CPC for maximizing the collection of radiation incident within a wedge (or cone) of half-angle $\theta_c$ onto a flat absorber is shown schematically in Figure 2.1.

The right branch of the concentrator (BC) is a segment of a parabola with its focus at point A, which lies on the axis of the full parabola indicated by the short dashed line so labeled. This axis of this parabola is tilted counterclockwise by an angle $\theta_c$ with respect to the desired CPC axis of symmetry. The left branch of the concentrator (AD) is the segment of a similar parabola corresponding to the image of BC obtained by reflection of BC through this symmetry axis. This parabola has its focus at point B, and its axis (not shown) is tilted clockwise by an angle $\theta_c$ with respect to the symmetry axis. The curved parabolic surfaces AD and BC are highly reflecting mirrors. Many details of this geometry and its characteristics were discussed extensively by Rabl (1976b).

CPCs can be either "two-dimensional," corresponding to cylindrical trough-shaped devices formed by translating the shape in Figure 2.1 perpendicular to the page, or "three-dimensional," corresponding to cone-shaped devices formed by rotating the shape around the axis of symmetry. In either case, the shape in Figure 2.1 is the cross-section profile of the corresponding device.

In two dimensions, it is easily seen by inspection of Figure 2.1 that

1. all rays crossing the entrance aperture DC at a counterclockwise angle $\theta$ equal to $\theta_c$ (i.e., parallel to the axis of the parabola) will strike the mirror BC and be brought to a focus at point A;
2. all rays making a counterclockwise angle $\theta \le \theta_c$ must be reflected toward points somewhere on the flat absorber AB; and
3. all rays making a counterclockwise angle $\theta \ge \theta_c$ will be reflected to points on the opposite mirror AD above point A and will be turned back and rejected.

Identical arguments hold for rays making clockwise angles with respect to the optical axis.

Furthermore, it is also easily shown from simple geometry that the ratio of the length of the entrance aperture DC to the absorber AB is $1/\sin\theta_c$ and that, therefore, the concentrator is said to be "ideal"—that is, the design achieves the maximum concentration given by Eq. (1.2a) and *all* the

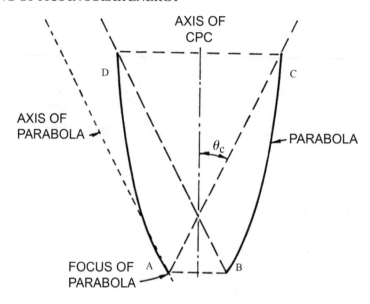

**Compound Parabolic Concentrator.**

**FIGURE 2.1:** Cross-section of a simple CPC designed for a flat absorber.

rays inside $\theta_c$ are accepted and *all* the rays outside $\theta_c$ are rejected (Collares-Pereira, O'Gallagher, and Rabl, 1978; Welford and Winston, 1978). That is all there is to it.

We only need two parameters to specify the complete geometry of a fully developed (i.e., untruncated) CPC: the width, $S$, of the flat absorber (AB in Figure 2.1) and the acceptance angle, $\theta_c$. Then, from Eq. (1.2a), with $n = 1$ and simple geometry, we find that the entrance aperture, $W$ (DC in the figure), and the untruncated height, $H$, of the CPC, respectively, are

$$W = S/\sin \theta_c \qquad (2.1)$$

and

$$H = S(1 + 1/\sin \theta_c)/(2\tan \theta_c) \qquad (2.2)$$

It is most convenient to use a parametric approach to calculate the coordinates of one branch or sidewall (say, the right branch AC in Figure 2.1) of a CPC in a useful coordinate system. We first determine the coordinates $x', y'$ in the "natural system" in which the $y'$ axis is identical with the parabola axis shown in the figure, and the origin lies on that axis such that the focus is at point A. The focal length $F$ of the parabola is

$$F = (S/2)(1 + \sin \theta_c) \qquad (2.3)$$

We express $y'$ in terms of $x'$ and vary $x'$ incrementally as an independent parameter over a range such as to specify the full sidewall (BC). We then rotate and translate the resulting set of $x',y'$ coordinates to provide a set of the $x,y$ coordinates in the desired coordinate system. To be more specific, we define the origin of the desired useful coordinate system to be the center of the flat absorber and the $y$ axis to be the axis of the CPC as shown in Figure 2.1. In this coordinate system, the apex of the parabola defining the right branch lies at

$$x_0 = -S/2 + F\sin\theta_c$$
$$= -(S/2)(1 - \sin\theta_c - (\sin\theta_c)^2) \qquad (2.4a)$$

$$y_0 = (S/2)(1 + \sin\theta_c)(\cos\theta_c) \qquad (2.4b)$$

To generate the set of coordinates for the right sidewall, we begin at point B in Figure 2.1, where

$$x_1' = S(\cos\theta_c) \qquad (2.5a)$$

and

$$y_1' = (S/2)(1 - \sin\theta_c) \qquad (2.5b)$$

We then use $x'$ as an incrementally variable independent parameter such that

$$x_{n+1}' = x_n' + \delta x' \qquad (2.6a)$$

and

$$y_{n+1}' = (x_{n+1}')^2/4F \qquad (2.6b)$$

with $\delta x'$ being some suitable small increment, such as $W/N$ where $N \sim {>}100$.

The right branch of the CPC sidewall profile in the desired coordinate system is then determined from Eqs. (2.5) and (2.6) with

$$x_n = x_n'(\cos\theta_c) - y_n'(\sin\theta_c) + x_0 \qquad (2.7a)$$

and

$$y_n(x_n) = x_n'(\sin\theta_c) + y_n'(\cos\theta_c) + y_0 \qquad (2.7b)$$

For the left branch, we have

$$y_n(-x_n) = y_n(+x_n) \qquad (2.7c)$$

The three-dimensional (cone) solution is identical in profile with the one given but rotated about the vertical axis.

In three dimensions (cone geometry), it is well known that, although the behavior of rays in the meridian plane (the plane containing the optical axis) is as described earlier, the corresponding devices are *not* ideal: That is, there are small losses of skew rays (rays not in the meridian plane) inside $\theta_c$ and, correspondingly, small acceptance outside $\theta_c$ of some skew rays. This behavior was discussed extensively by Welford and Winston (1978). However, despite this "nonideality," these cone concentrators have the geometric concentration given by Eq. (1.2b), and the "skew-ray losses" occur for incident angles close to $\theta_c$ and are generally small; therefore, as a practical matter, such cone CPCs are very effective.

One of the most interesting discoveries in the early days of nonimaging optics was the relevance of these concepts to evolutionary biology. In particular, it was noted that the retinal cone receptors in the eye of the horseshoe crab, which evolved more than 300 million years ago, turn out to be almost perfect CPC cones (Levi-Setti, Park, and Winston, 1975).

## 2.2    "TRUNCATION" OF CPCs

One important practical consideration is the desire to reduce the depth of the concentrator and corresponding mirror area. Because the portions of the CPC that are nearly parallel the optical axis are not contributing much to the size of the entrance aperture, most CPCs are cut off or "truncated" well short of full development. As a rule of thumb, truncation to about half the fully developed height yields a good tradeoff between concentration and mirror area. This issue was discussed extensively by Rabl (1976b).

## 2.3    THE "EDGE-RAY PRINCIPLE"

The basic "tilted parabola" CPC for a flat absorber as shown in Figure 2.1 is the simplest illustration of a broader principle that allows these techniques to be generalized to cases with absorbers of different shapes for so long as the shape has no concave curvature. This is the fundamental principle of nonimaging optics and is referred to as the "edge-ray principle" (Welford and Winston, 1989; see also Ries and Rabl, 1994). For our purposes, it can be simply stated as follows:

*To maximize the collecting aperture for radiation onto a generalized absorber shape in two dimensions within a given field of view defined by a pair of extreme rays, one simply requires that the slope of the optical surfaces be such as to redirect the extreme rays so that they are tangent to the edge of the desired absorber.*

An equivalent alternative statement of this principle (Welford and Winston, 1980) is

*Maximum concentration is achieved by ensuring that rays collected at the extreme angle for which the concentrator is designed are redirected, after at most one reflection, to form a caustic on the surface of an absorber.*

Starting with a specified nonconcave absorber shape and one of the extreme rays ("edge rays"), application of this principle yields a differential equation for one branch of the reflector shape that will maximize collection in two dimensions. Application of this principle with the other extreme ray will generate the second branch. Note in particular that the two extreme rays need not be symmetric about a reference axis (i.e., $\theta_c+$ in the clockwise direction need not be equal to $\theta_c-$ in the counterclockwise direction), thus leading to asymmetric CPCs, if so desired (Rabl, 1976a; Mills and Giutronich, 1978).

## 2.4    SOLUTIONS FOR NONFLAT ABSORBERS

Symmetric two-dimensional (trough-like) CPCs for three new absorber shapes, in addition to the flat absorber case (Figure 2.1) already discussed in detail, are shown in Figure 2.2. Figure 2.2b shows a vertical fin absorber in which the extreme rays are directed to be tangent to the edge of the fin at its top. It is easily seen that this solution consists of two tilted parabolic segments, each with its focus at the top edge of the fin. Inside the shadow lines, the reflector shape is simply an involute of the absorber shape. In this case, the involute of a fin is a circle of radius equal to the height of the fin, with its center at the top edge of the fin. Figure 2.2c is for a wedge absorber. The reflector again consists of two tilted parabolas, each with its focus at the top of the wedge. The sidewall profile coordinates for the parabolic segments of the fin and wedge solutions are easily generated by a parametric procedure similar to that described in detail for the flat absorber above. The only difference is in the relative placements of the foci and origins in the two coordinate systems.

Finally, Figure 2.2d shows the shape for a round circular absorber. This shape is of the most practical interest because it is the two-dimensional solution for a tubular absorber of circular cross-section. The shape is not a true parabola at all because application of the "edge-ray principle" yields a shape that reflects the extreme ray at each point to a different tangent point on the absorber circumference. Thus, there is no unique focal point.

The solution, first presented by Winston and Hinterberger (1975), and the mathematical description of the shape are best understood in terms of a parametric relationship described in detail

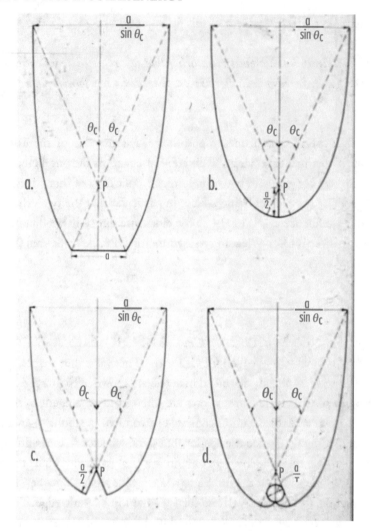

**FIGURE 2.2:** Two-dimensional (trough-shaped) CPC reflectors for four absorber configurations as described in the text. The flat absorber (a) and the circular (tubular) absorber (d) shapes are the most useful practical shapes.

by Rabl (1976c). If the origin of the coordinate system is taken as the center of the circular absorber tube of radius $a$, it is convenient to express the geometry of the sidewall profile terms of the coordinates, $\rho$ and $\theta$, as defined in Figure 2.3. Here, $\rho$ is the length of the tangent line from the point of tangency to the reflector and $\theta$ is the parametric arc angle of the point of tangency relative to the downward vertical direction (i.e., the bottom of the circle). Inside the "shadow lines" ($\theta < \theta_c + \pi/2$), the coordinate $\rho$ can also be visualized as the length of a taut string being "unwrapped" from around the circular absorber and $\theta$ is the "unwrapping angle." The coordinates of the reflector inside the

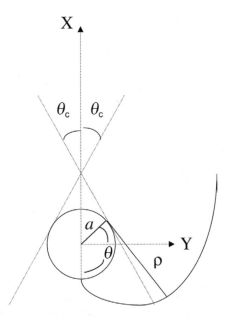

**FIGURE 2.3:** Geometry defining the parameters used in the parametric equations to calculate the actual CPC surfaces for circular (tubular) absorbers (see Rabl, 1976c).

"shadow lines" defined by the extreme rays tangent to the absorber at the acceptance angle are those of a simple involute of the absorber cross-section beginning at the bottom of the circle,

$$\rho = a\theta \qquad (2.8)$$

Outside the shadow lines $(\theta > \theta_c + \pi/2)$, the Winston–Hinterberger solution is given by

$$\rho = a(\theta + \theta_c + \pi/2 - \cos(\theta - \theta_c)/(1 + \sin(\theta - \theta_c)) \qquad (2.9)$$

Both inside and outside the shadow lines in Figure 2.3, the $x$–$y$ coordinates of the right-hand sidewall profile are then

$$x = a \sin \theta - \rho \cos \theta$$
$$y = -a \cos \theta - \rho \sin \theta \qquad (2.10)$$

The coordinates for a fully developed (untruncated) CPC of acceptance angle $\theta_c$ for a circular absorber of radius $a$ are then generated by applying Eqs. (2.8), (2.9), and (2.10), starting at $\theta = 0$ and incrementing by suitably small $\delta\theta$ values for all $\theta$ values within $0 \le \theta \le \pi + \theta_c$.

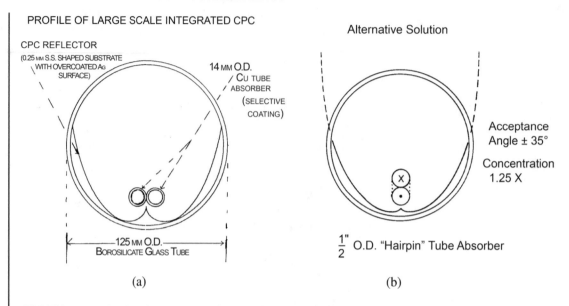

**FIGURE 2.4:** CPC reflector shapes ($\theta_c = \pm 35°$) for special "figure 8" absorber cross-section corresponding to a "hairpin"-shaped U-tube absorber tube. (a) Horizontal orientation that eliminates gap losses; (b) vertical orientation.

Many other two-dimensional solutions are possible, and it is beyond the scope of this monograph to present them all. However, two additional solutions of practical interest for "hairpin"-shaped U-tube configurations are shown in Figure 2.4a and 2.4b. It should be emphasized in these cases that the reflector is developed for a virtual absorber (shown in the figure by dotted segments) because the real absorber has sections that are concave. Thus, the actual geometric concentration for the real absorber is slightly less than ideal. (The real absorber surface is larger than the virtual surface used with the edge-ray principle.)

## 2.5   THE "GAP-LOSS" PROBLEM

It can be shown (Winston and Welford, 1978; Welford and Winston, 1989) that for a truly ideal reflecting two-dimensional cylindrical concentrator to achieve the thermodynamic limit (Eq. (1.2a)) without geometric losses, the reflector must touch the absorber. Since this is undesirable and/or impractical in most thermal collectors (to allow for mechanical tolerances and to prevent the reflector from providing a heat loss path from the absorber to ambient), a small separating "gap" is usually designed into the concentrator. This gap results in a small number of rays being lost (i.e., reflected back out of the concentrator without hitting the absorber)—how small depends on the size of the gap relative to the absorber perimeter. Such a "gap" is obviously necessary if the absorber is to be surrounded by an evacuated glass envelope.

Much work has gone into devising methods of reducing or eliminating such losses. See, for example, the studies by Winston (1978), Rabl, Goodman, and Winston (1979), and Welford and Winston (1989). One approach to a practical collector design in such cases has been simply to calculate the reflector profile for the given absorber but to terminate the reflectors at the point at which they intersect the obstruction. It can be shown (Rabl, Goodman, and Winston, 1979) that, if this is done, the resulting gap loss averaged over the acceptance angle will be given by

$$L = (1/\pi)[\{2g/r + (g/r)^2\}^{1/2} - \text{arc } \cos\{r/(g + r)\}]] \qquad (2.11)$$

Hence, the optical losses can become quite large if the gap size is comparable with the absorber dimensions. For small gaps, $g \ll r$, $L$ is well approximated by

$$L \cong g/\pi r \qquad (2.12)$$

Winston developed several more sophisticated solutions that are considered to be quite effective. We mention only two of those here: (1) the "extended cusp" solution and (2) the "compromise gap loss suppression" solution.

The extended cusp gap solution (Winston, 1978) is illustrated by the mirror profile curve shown in Figure 2.5 for a case with a gap $g = r_2 - r_1$ equal to the absorber radius itself. Terminating the standard CPC solution at its intersection with the outer circle leaves a gap, which results in a gap loss of ~22% (from Eq. (2.11)); thus, the actual energy reaching the absorber even with otherwise perfect optics will only be ~78% of what it could be for the design acceptance angle. The extended cusp solution is generated by constructing a "virtual absorber" by laying out tangent lines to the absorber from point D (the point along the extension of the desired optical axis where it meets the obstruction). The reflector curves then begin at point D and correspond to the ideal solution obtained by applying the fundamental edge-ray principle to the virtual absorber (DBAA'B'D). The practical implementation of this solution is simply to start the mathematical involute curve (of Eq. (2.8)) from a different point on the absorber: That is, we start "unwrapping" string from a different arc angle on the absorber—the angle (clockwise with respect to the bottom of the circle) where the end of the string stretched from point B to point D in Figure 2.5 would lie if this string were to be wrapped onto the absorber. Of course the actual physical reflectors in this case exist only outside $r_2$. This new solution will accept no radiation outside the acceptance, but it is not an ideal concentrator in the usual sense because it will still reject some rays entering the aperture within the acceptance angle but passing through the gap. The fraction of the rays that are collected can be shown to be the ratio of the actual absorber circumference ($2\pi r_1$) to the full perimeter of the virtual absorber. The

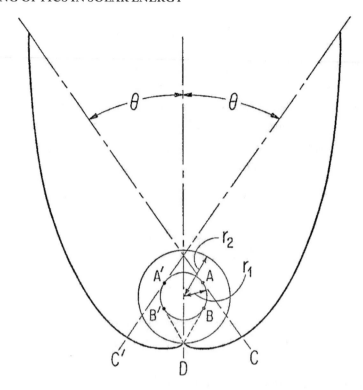

**FIGURE 2.5:** Cross-section profile of the "extended cusp" solution for a nonimaging concentrator designed for a tubular absorber with reflector gaps. The solution is the Winston–Hinterberger solution for the "ice cream cone" absorber DB′A′ABD.

corresponding losses as a fraction of the aperture area (about 18% for $r_2 - r_1 = g = r_1$) are not reduced dramatically below that of the conventional solution, but the extended cusp solution has an aperture $A_1 > \pi r_1/\sin\theta$ such that the net flux concentration seen by the absorber is $1/\sin\theta$ and is therefore the maximum permitted for this acceptance angle. It should be emphasized that this improved design attains both of the usual advantages for which concentration is used: That is, both thermal performance and cost reduction have to do with getting as much energy onto the receiver as possible. In our example here, the total amount of energy within the acceptance angle is increased from roughly 4/5 of the thermodynamic maximum to the maximum allowed—a relative increase of about 25%.

The compromise gap solution is somewhat similar to the modified cusp just discussed in that one designs for a virtual absorber configured so as to reduce but not entirely eliminate the gap losses. Here, the beginning arc angle of the involute lies in the other direction (counterclockwise with respect to the bottom of the circle) from that for the extended cusp solution. This introduces some sacrifice in concentration (the virtual absorber is smaller than the real absorber). The reflectors thus begin outside $r_2$, and the bottom of the absorber tube is viewed by a "cavity" formed by one or (sometimes but rarely) more "Vee grooves," which have the effect of increasing the throughput to compensate for some or most of the losses in the extended cusp solution (see O'Gallagher, Rabl, Winston, and McIntire, 1980). These matters are discussed in more detail in by Welford and Winston (1989).

## 2.6    CPC SOLAR GEOMETRY

The fundamental improvement provided by nonimaging optics in general and CPCs in particular is an increase in the field of view for a given geometric concentration. With CPCs, this allows useful concentration to be achieved without active tracking. There are several ways to do this; the most common is illustrated in Figure 2.6. The long axis of the CPC troughs is aligned in an east–west direction, and that of the normal to the trough apertures is tilted downward from the zenith by an angle equal to the latitude angle. The angular acceptance is a wedge of half-angle $\pm\theta_c$. This wedge traces out an "orange slice" on the celestial sphere, and whenever the sun's path lies within this orange slice, all of the direct solar radiation is collected, concentrated, and delivered to the absorber.

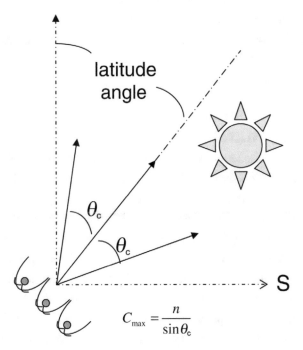

$$C_{max} = \frac{n}{\sin\theta_c}$$

**FIGURE 2.6:** The basic deployment geometry for an east–west-aligned CPC in the northern hemisphere is illustrated (as viewed looking east).

For any value of $\theta_c$, on the equinoxes, the sun's path is at the center of the field of view all day. During spring and summer, the sun's path lies in the upper half of the acceptance; during fall and winter, in the lower half. With $\theta_c = 35°$, the maximum concentration $C_{geom}$ (for an untruncated CPC) is 1.74× and the sun lies within the acceptance for 7 hours on the solstices (both winter and summer) and for longer periods the rest of the time. The collection can be biased toward summer (or winter), if desired, giving longer collection times during this season. Higher concentrations require smaller values of $\theta_c$ and therefore reduced collection times or some seasonal adjustment of the tilt angle. For instance, a concentration of 3× can be achieved with $\theta_c = 18°$ and two tilt orientations (one for summer and another for winter). Practical concentrations up to ~10× can be achieved with about 12 tilt adjustments a year. Concentrations greater than 10× would require some movement during the day in order to maintain acceptable collecting times. Furthermore, at these larger concentration ratios ($\geq 10×$), the height-to-aperture and reflector area-to-aperture area ratios become prohibitive. For a comprehensive discussion of these issues, see the work of Rabl (1976).

Far and away, the most common CPC is the $\theta_c = 35°$, $C_{geom} \cong 1.3×-1.6×$, completely stationary concentrator that requires no movement whatsoever. Another configuration is a design with an even lower concentration (1.1×–1.2×) and a larger acceptance angle ($\theta_c = 55°$) that allows stationary concentration with either a horizontal (east–west) or a polar (north–south) orientation. In the polar orientation, such collectors "see" the sun for about 7 1/3 hours a day all year.

It should be emphasized that all these low-concentration designs effectively "unwrap" and magnify the entire perimeter of the absorber cross-section. Thus, the aperture of a 1.5× CPC is 1.5 times $\pi$, or 4.7 times the tube diameter.

## 2.7    CPC DEPLOYMENT FLEXIBILITY

Since the CPC does have limited angular acceptance, it should be mounted as close as possible to the nominal design configuration shown in Figure 2.5. However, considerable flexibility is possible. Computer modeling has shown quantitatively that the effects of nonstandard deployment orientations on the long-term projected collected energy of CPCs may be quite acceptable over a relatively large range of directions. For example, we analyzed the energy collected by a ±35° CPC in Phoenix, Arizona, as calculated from hour-by-hour simulations based on typical meteorological year data. The results are shown in Figure 2.7 as a function of the misalignment angle for three orientations of the misalignment axis: (1) perpendicular to the collector aperture; (2) parallel with the ground and pointing due south; and (3) perpendicular to the ground (i.e., vertical). The geometric effects were found to be similar for all three orientations. The important conclusions are that for deviations of ±10°, the dropoff compared with the ideal orientation is less than 3%, is about 5% at ±20°, and remains less than 10% out to ±30°. Similar behavior is found at other locations, such as Boston and

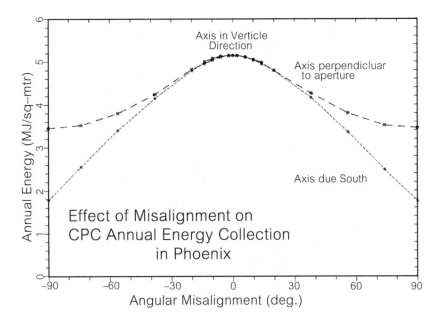

**FIGURE 2.7:** Energy collected by a misaligned (with respect to due south at tilt equals latitude) ±35° CPC in Phoenix, Arizona, as calculated from hour-by-hour simulations based on typical meteorological year data. Results are shown for three orientations of the misalignment axis: (1) perpendicular to the collector aperture; (2) parallel with the ground and pointing due south; and (3) perpendicular to the ground (i.e., vertical). The results for cases 2 and 3 are virtually indistinguishable.

Miami, such that, in general, it can be concluded that, although they are more sensitive to alignment than flat plates, CPC collectors retain a considerable flexibility in deployment.

## 2.8  THE DIELECTRIC TOTALLY INTERNALLY REFLECTING CPC

We cannot leave this topic without discussing another nonimaging device with practical applications for PV application in both single-stage and multistage concentrators. This is the solid transparent CPC made of refracting or dielectric material and designed such that total internal reflection occurs for all rays within the design acceptance angle. Such devices are usually referred to with the somewhat cumbersome designation as *dielectric totally internally reflecting CPCs* (DTIRCs). Their basic design and operation are shown in Figure 2.8. Consider a solid transparent CPC of a given shape (and geometric concentration ratio): Essentially, for rays in the meridian plane, refraction at the flat entry aperture allows an increase in the sine of the acceptance angle by a factor of $n$ (the

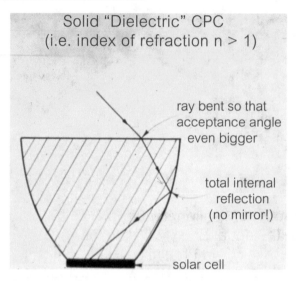

FIGURE 2.8: Meridian plane cross-section of a DTIRC.

index of refraction of the material), in accord with Eq. (1.2a). Furthermore, it turns out that one can usually design the sidewall profile such that the condition for total internal reflection for all rays within the acceptance angle is satisfied (Winston, 1976) so that no mirror is required! An additional benefit of this approach occurs for nonmeridianol rays (rays that are not in a plane perpendicular to the long axis of the trough. For such rays, the effect of Snell's law is to widen the acceptance angle further as the longitudinal projected angle increases. This lengthens the collecting day for a solid transparent refracting trough. In particular, it means that a concentration ratio of 4× can be achieved with a totally stationary concentrator.

Finally, note that by curving the front surface, it can be made to act as a lens so that the overall height for the same concentration can be decreased (Ning, Winston, and O'Gallagher, 1987a). (See further discussions of these devices as secondary concentrators in Chapters 6 and 9.) These devices are not effective for most thermal applications but do have particular advantages for PV applications.

## 2.9    SUMMARY OF CPC FEATURES

The CPC is a straightforward example of an ideal light collector. In its two-dimensional and trough-shaped implementation, it has many advantages for solar concentration.

In particular,

1. CPCs achieve the widest possible angular field of view for a given geometric concentration.
2. This permits useful concentrations without tracking:

    (a)  1.1×–2× for totally stationary collector and

    (b)  2×–10× with seasonal adjustment.

   3.   CPCs collect a large fraction of diffuse components of sunlight.

Higher concentrations (>10×–>40,000×) are usually achieved with multistage systems in both two-dimensional (line focus) and three-dimensional (point focus) geometries. These systems do require some tracking but allow use of relaxed optical and tracking tolerances.

C H A P T E R  3

# Practical Design of CPC Thermal Collectors

As we have noted, the major goal of employing concentration in solar thermal collectors is to improve performance by reducing heat losses. This is particularly effective when the heat losses are already quite low, such as for absorbers with vacuum insulation and a spectrally selective coating (Garrison, 1979a, b; O'Gallagher and Winston, 1983). The vacuum, as long as it is less than ~$10^{-6}$ Torr ($1.33 \times 10^{-4}$ Pa), essentially eliminates convection and conduction losses so that the only losses are radiative. The selective coating with an emittance in the infrared of 0.1 or less greatly suppresses these even further, so that the relatively low concentrations attainable with stationary CPCs result in a collector with superb high-temperature efficiency for a stationary nontracking collector. The main features of this approach are discussed in the first part of this chapter. The advantage of concentration is not so pronounced if the absorber is not evacuated; however, there remain both performance and cost advantages for nonevacuated CPC thermal collectors, and these are reviewed in the second portion of this chapter.

## 3.1 CPCs WITH EVACUATED ABSORBERS
### 3.1.1 The External Reflector CPC
The CPC has been under development for solar energy for more than 30 years (Welford and Winston, 1978; O'Gallagher and Winston, 1983). It has always held out the promise of being able to provide efficient conversion of sunlight to thermal energies at temperatures well above 200 °C without tracking. The first generation of evacuated CPC collector designs was developed at the Argonne National Laboratory in the early 1970s. The basic design involved coupling a truncated CPC-shaped reflector (Rabl, 1976a; McIntire, 1979; O'Gallagher and Winston, 1983) to an all-glass, dewar-type solar absorber tube (Figure 3.1). Such a tube was developed in the United States during the 1970s; by the latter part of that decade, evacuated solar collectors with external reflector CPC (XCPC)-type nonimaging concentrations were available from three U.S. manufacturers. The CPC acceptance angle is chosen to be approximately $\pm35°$, sufficient to allow daylong collection

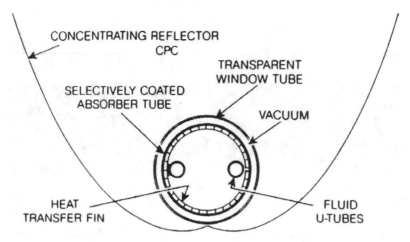

**FIGURE 3.1:** Cross-sectional profile of a basic stationary XCPC trough coupled to a dewar-type evacuated tube according to the design developed by the Argonne National Laboratory in the early 1970s. In addition, a cover glass across the aperture (not shown) is usually (but not always) used.

throughout the year with a fixed, totally stationary concentrator. The corresponding maximum concentration ratio is about 1.7×, but the reflector is usually truncated such that the actual concentration is between 1.1× and 1.5×.

We use the early Argonne National Laboratory XCPC design as the context with which to present the main features of this type of collector to illustrate the design concepts and to develop an optical and thermal model. We then review a more recent embodiment of the concept developed in the late 1990s (O'Gallagher, Winston, Mahoney, Dudley, Luick, and Gee, 2000, 2001) to see how the performance and potential lower costs have evolved over the 30-year period of development. This most recent embodiment of the concept was motivated by both economic and technical developments that promise to reduce costs and improve performance.

**3.1.1.1   Generic Model for Nonimaging Evacuated Solar Thermal Collectors.** The operating efficiency of a solar collector is defined as the ratio of the usable thermal power delivered to the solar radiant power incident on the collector aperture; that is,

$$\eta = P_{useful}/P_{incident} \qquad\qquad (3.1)$$

Here, $P_{incident} = A^*I$, where $A$ is the aperture area and $I$ is the incident hemispherical solar insolation.

The net useful thermal power, $P_{useful}$ is the fraction of the incident energy that makes it to the heat transfer fluid less any thermal losses. The optical losses include geometric losses as well as absorption and reflection losses. These are usually all lumped together in terms of the net optical efficiency $\eta_o$. Basically, this is the product of the transmittance of the cover glazing, a factor incorporating the effect of reflection losses (if any), any geometric losses, and the absorptance of the absorber surface. It is sometimes customary to include the so-called heat removal factor $F'$ (Duffie and Beckman, 1991) in this factor as well.

Besides optical loss and first-order heat removal losses, if the average operating temperature of the collector is higher than the ambient temperature, there will be thermal losses as well, such that the net useful power is given by

$$P_{useful} = \eta_o AI - P_{Loss}(T_{coll}, T_{amb})^* A \qquad (3.2)$$

where $P_{Loss}(T_{coll}, T_{amb})$ is heat loss per unit area per unit time, which is a function of $T_{coll}$ and $T_{amb}$, the average collector operating temperature and the ambient temperature, respectively. Thus, the operating efficiency as a function of temperature and insolation can be expressed as

$$\eta(T_{coll}, T_{amb}) = \eta_o - P_{Loss}(T_{coll}, T_{amb})/I \qquad (3.3)$$

**Simple model for XCPC optical performance.** The optical performance of a CPC collector is usually characterized simply by its "optical efficiency," $\eta_o$, and its angular acceptance, $\theta_c$. Here, $\eta_o$ is intended to represent the fraction of solar radiation incident on the aperture, which is absorbed by the heat transfer fluid before taking into account the effect of any thermal losses. Of course, in practice, the optical throughput will be a complicated function of incidence angle that can be best understood by detailed ray tracing. However, the approximate value of $\eta_o$ is a useful parameter that can be roughly estimated from the performance parameters of the materials used in the collector.

For incident sun angles within the design acceptance angle, the optical efficiency, $\eta_o$, for a glazed CPC with external reflectors is given by

$$\eta_o = \tau_1 \tau_2 \rho^{\langle n \rangle} \alpha (1 - L) \Gamma \qquad (3.4)$$

where $\tau_1$ and $\tau_2$ are, respectively, the transmittances of the cover glass and the outer glass vacuum envelope, $\rho$ is the reflectivity of the CPC mirror surface, $\langle n \rangle$ is the average number of reflections (typically slightly less than unity), $\alpha$ is the absorptance of the selective surface, $L$ is any geometric loss due to the "gap" between the reflector and the absorber surface, and $\Gamma$ is the fraction of the hemispherical incident solar radiation accepted by the CPC after correction for some "loss of diffuse." Here, we are treating the heat removal factor $F'$ as being near unity, which is often the case for evacuated tubes where the heat transfer surface is also a hydronic interface. However, for some cases, as discussed subsequently, this factor can become quite significant.

A detailed understanding of the optical properties of any collector including gap losses and cavity enhancement effects requires a comprehensive ray trace. Such an analysis would be carried out in the course of the design of a new collector, but its presentation here is beyond the scope of this monograph.

**Concentration and thermal performance for evacuated CPCs.** We model the thermal performance of our prototype collector panel using a simple parameterization of the heat loss per unit area per unit time, $P_{Loss}$, for an evacuated tube collector operating at an absorber temperature, $T_{abs}$, in an ambient temperature, $T_{amb}$, as follows:

$$P_{Loss}(T_{abs}, T_{man}, T_{amb}) = U_M(T_{man} - T_{amb}) + \varepsilon\sigma(T_{abs}{}^4 - T_{amb}{}^4)/C \qquad (3.5)$$

where we have included the losses from an insulated but nonevacuated manifold at temperature $T_{man}$, characterized by a linear heat loss coefficient per unit collector area, $U_M$, and $\varepsilon$, $\sigma$, and $C$ are the absorber surface emittance, the Stefan–Boltzmann constant [$5.67 \times 10^8$ W/(m²-K)], and the geometric concentration of the CPC reflector, respectively. The thermal performance, which is the solar-to-thermal conversion efficiency, $\eta$, will then be given by

$$\eta(T_{abs}, T_{man}, T_{amb}) = \eta_o - U_M(T_{man} - T_{amb}) + \varepsilon\sigma(T_{abs}{}^4 - T_{amb}{}^4)/CI \qquad (3.6)$$

where $I$ is the hemispherical solar insolation incident on the collector aperture and $\eta_o$ is the optical efficiency of the collector. Note that the radiative contributions to the thermal loss scale are the reciprocal of the geometric concentration ratio. For an absorber with already low heat losses (e.g., an evacuated selective absorber), even very low levels of concentration can be very effective. This is illustrated in Figure 3.2 for operating temperatures up to around 300 °C and shows clearly how, in this application, the first few factors of 2 or so are much more effective at these temperatures than very high concentrations. This is the reason that it is felt that some form of evacuated selective nontracking CPC solar collector has the potential to be the general-purpose solar thermal collector of the future.

Almost all experimental and commercial developments of evacuated CPCs over the last two decades have emphasized the stationary version of the collector—that is, one having a large acceptance angle (between 35° and 50°) and a geometric concentration ratio between 1.0× and 1.6×. This reflects the perceived major advantage of a totally fixed collector over one that requires any adjustment at all—even if that is only a few times per year. It is often said that "if it has to move at all, you may as well track." There may, however, be many applications in which this is not the case—in which, in fact, the additional concentration provides such a significant performance boost that it more than offsets the costs of a seasonal adjustment (Figure 3.2). It should be recalled that

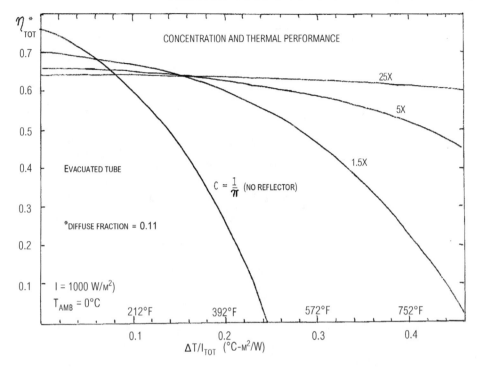

**FIGURE 3.2:** Illustration of the effect of introducing low levels of concentration in an evacuated tubular solar collector. The first few factors of 2 are the most effective for temperatures up to about 300 °C.

evacuated collectors in the intermediate concentration range were developed and tested in the early days of nonimaging optics and demonstrated remarkable thermal performance at temperatures approaching 300°C. A prototype 5× CPC for an evacuated dewar-type tubular absorber was built and tested at the University of Chicago (Collares-Pereira, O'Gallagher, Rabl, and Winston, 1978) and a 5× CPC for an evacuated flat fin absorber was designed at the Argonne National Laboratory. Both of these collectors required only 12 adjustments per year, and both of these concepts could provide a competitive alternative to tracking parabolic troughs for solar thermal electricity generation.

## 3.1.2   Comparison of Early and Recent XCPC Embodiments

The cross-sectional profile of the most recent experimental embodiment of the XCPC concept is shown in Figure 3.3. The design was motivated by both economic and technical developments that promised to reduce costs and improve performance (O'Gallagher et al., 2000, 2001).

Values of the various parameters are listed in Table 3.1 for the original Argonne National Laboratory prototype from the 1970s along with those for the more recent experimental prototype version from the 1990s (Figure 3.1). For the early versions of this type of collector, $\tau_1$ and $\tau_2$ were

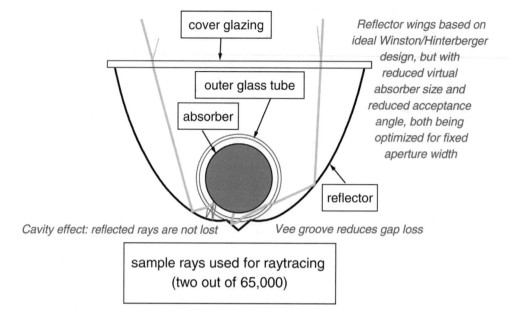

FIGURE 3.3: Cross-section profile of a more recent version of an XCPC collector trough. The optimized nonimaging reflector profile is designed for an acceptance half-angle of 30° and couples an inexpensive evacuated dewar-type absorber to an aperture with a geometric concentration ratio of 1.2×. The optical design was the result of a comprehensive optimization procedure. This design has the potential to provide a low-cost path to a nontracking high-temperature (>200 °C) collector.

~0.91, $\alpha = 0.82$, $\rho = 0.88$, $\langle n \rangle \cong 0.8$, $L$ was ~0.02, and $\Gamma$ was ~0.98. Taking the product of all these factors together yields an expected value for the optical efficiency of $\eta_o \cong 0.59$. This is the efficiency for insolation at normal incidence. At other nonnormal incidence angles, this factor must be multiplied by the transverse and longitudinal incidence angle modifiers (IAMs), which incorporate the angular dependence of the optical characteristics of the concentrator and the angular properties of the reflection, transmission, and absorption coefficients.

For the 1990s version, as can be seen from Table 3.1, the major improvements are an increase in reflectance to about 0.95, an increase in absorptance to about 0.90, and an improved optical design to reduce the gap losses. Here, $L$ can be estimated from radiation transfer methods and is found to be about 0.5%. As a rule of thumb, an ideal low-concentration collector "sees" about $1/C$ of the diffuse insolation. This will be reduced slightly to the extent that cavity effects are present and increased if the diffuse radiation is not isotropic but is incident preferentially from directions near the sun. If one assumes a typical diffuse fraction of 12%, the foregoing considerations lead to an estimated value of $\Gamma = 0.98$ for $C = 1.2×$. Taking the product of all factors together yields an

**TABLE 3.1:** Performance Parameters for XCPC Designs

| PARAMETER | SYMBOL | 1970S (ARGONNE NATIONAL LABORATORY) VERSION | 1990S VERSION |
|---|---|---|---|
| Transmittance of cover glass | $\tau_1$ | 0.91 | 0.91 |
| Reflectivity of mirror surface (Alcoa Everbrite) | $\rho$ | 0.88 | 0.96 |
| Average number of reflections | $\langle n \rangle$ | 0.8 | 0.8 |
| Transmittance of outer tube | $\tau_2$ | 0.91 | 0.91 |
| Absorptance of selective surface (normal incidence) | $\alpha$ | 0.82 | 0.90 |
| Gap losses | $L$ | 0.02 | 0.005 |
| Correction for loss of diffuse | $\Gamma$ | 0.98 | 0.98 |
| Net optical efficiency (Eq. (3.1)) | $\eta_o$ | 0.59 | 0.70 |

expected value for the optical efficiency of the 1990s prototype of $\eta_o \cong 0.70$. A comprehensive ray trace analysis predicts that this value may be enhanced by several percentage points due to cavity effects (O'Gallagher, Rabl, Winston, and McIntire, 1980).

The measured optical efficiency (12) is $\eta_o = 0.68 \pm 0.035$, which is in good agreement with the optical properties of the collector tube, reflector surface, and cover glass transmittance.

The value of the geometric concentration $C$ for this XCPC is 1.2. Based on previous experience with integrated evacuated CPCs (ref), we estimated that a value of $U_M = 0.2$ W/(m²-K) would be readily achievable. Although the manufacturer's specifications list a value of $\varepsilon = 0.06$ for the emittance at room temperature for the absorber surface, we determined from stagnation measurements that a value of $\varepsilon = 0.12$ at high temperatures (approaching ~200 °C) is more representative.

The thermal performance for this CPC predicted by our model is compared with measurements in Figure 3.4 (O'Gallagher, Winston, Mahoney, Dudley, Luick, and Gee, 2001). The measured performance remains very respectable at temperatures between 100 °C and 200 °C compared with what can be achieved by a nonevacuated flat plate collector and even compared with evacuated

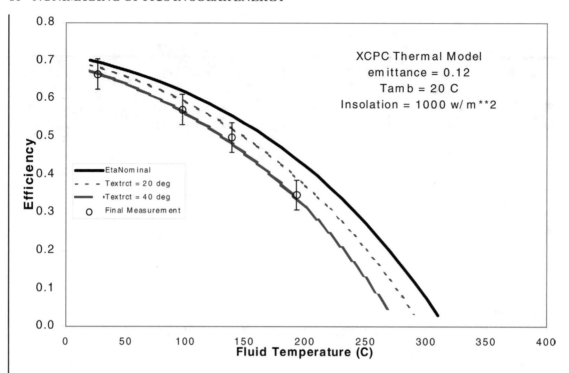

**FIGURE 3.4:** Thermal performance of XCPCs predicted by the simple model described in the text for three assumptions regarding the effectiveness of the heat extraction fin (O'Gallagher, Winston, Mahoney, Dudley, Luick, and Gee, 2001). The open circles represent the measured efficiency of the panel after modifications to reduce manifold losses and are consistent with other evidence that the heat extraction can still be improved.

tubular collectors that do not employ concentration. However, it is clear that the actual performance is still somewhat below the predicted performance. We interpret this as being due to relatively poor heat extraction in the dewar-type evacuated tube. If the heat transfer is not effective, the absorber surface may be at a somewhat higher temperature than that of the heat transfer fluid and the thermal performance may be less than optimal. To first order, we could define the heat removal factor to be less than unity by a significant amount. However, properly incorporating this effect for an evacuated tube requires a model more sophisticated than a simple constant correction factor. In our model, this effect is characterized by $T_{ext}$, the difference between the absorber temperature, $T_{abs}$, and the manifold temperature, $T_{man}$ (assumed to be the same as the operating fluid temperature): That is, in Eq. (3.6), we set $T_{abs} = T_{man} + T_{ext}$. Further development is expected to be able to improve this heat extraction so that the performance can be increased even further.

### 3.1.3   The Integrated CPC

We have seen how the first versions of evacuated CPCs, developed at the Argonne National Laboratory, used XCPCs coupled to evacuated dewar-type absorbers and led to commercial collectors manufactured by several manufacturers. Furthermore we have seen how this path had evolved over the past 30 years. However, soon after the early XCPC designs were developed, it became clear that considerable further improvement in performance and potential manufacturability could be had by integrating the reflector and absorber into an evacuated tubular module (Garrison, 1976a, b) The University of Chicago, in collaboration with GTE Laboratories, began an intensive program to develop an advanced evacuated solar collector integrating CPC optics into the design. An experimental preprototype panel of these "integrated CPC" (ICPC) tubes achieved the highest operating efficiency at high temperatures ever measured with a nontracking stationary solar collector (Snail, O'Gallagher, and Winston, 1984) as shown in Figure 3.5. Details of the optimum materials and performance parameters are given in a great number of previous reports (Snail et al., 1984; Winston,

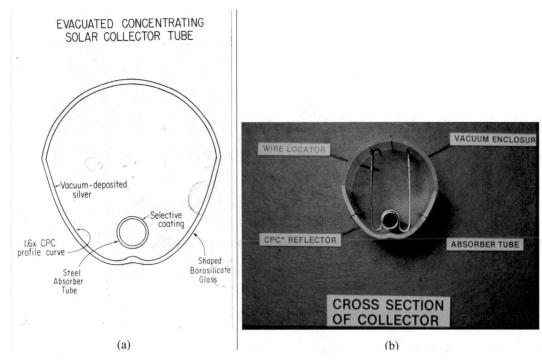

(a)                                            (b)

**FIGURE 3.5:** (a) Schematic illustration of an ICPC in which the optics is incorporated directly into the vacuum envelope. Here, the CPC profile is achieved by shaping the glass tube and the reflecting surface is front surface silvered. (b) Photograph of a cutaway sectional profile of an ICPC collector.

O'Gallagher, Muschaweck, Mahoney, and Dudley, 1999a, b). Arrays of such ICPC tubes have the potential to become the general-purpose solar thermal collectors of the future.

The ICPC prototype profile shown in Figure 3.6 has a full angle of acceptance of 70° and is oriented with its long axis in the east–west direction and deployed such that the plane of its aperture is tilted by an angle equal to the local latitude.

We believe that some form of advanced ICPC is the only simple and effective method for delivering solar thermal energy efficiently in the temperature range of 100 °C to about 300 °C without tracking. It can provide an efficient source of solar heat at these temperatures and makes practical and economical several cooling technologies that are otherwise not viable. When driven by these collectors, such advanced cooling technologies as double-effect or multistage regenerative absorption cycle chillers and high-temperature desiccant systems will have significantly improved performance and can achieve an overall system coefficient of performance high enough to be economical in a wide variety of applications. Even the best flat plate collectors cannot deliver the required high temperatures and thermal performance. Furthermore, tracking parabolic troughs are not likely to be practical in residential-scale systems or effective in many environments where cooling is desired.

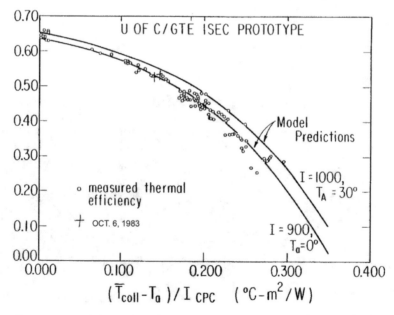

**FIGURE 3.6:** Comparison of the predicted model performance and the actual measured thermal performance of a panel of ICPC tubes of the design shown in cross-section in Figure 3.5. This collector achieved the highest instantaneous efficiency at operating temperatures higher than 250 °C ever achieved by a collector designed to be completely nontracking and that is fully stationary throughout the year.

The ICPC is the only high-temperature nontracking option available. In addition to its potential for driving cooling systems, this technology also provides a highly versatile solar source for virtually all thermal end uses, including general-purpose space and domestic hot water heating, as well as industrial process heat.

**3.1.3.1 A "Manufacturable Design".** Market conditions and remaining technical barriers to high-volume manufacturing have slowed progress and delayed the transfer of this very promising technology to industry and the commercial sector in the United States. Recent studies have indicated that if a mass-producible collector incorporating the same basic concepts as our prototype can be developed, it will be able to deliver cost-effective high-temperature performance across a broad range of temperatures, extending from about 50 °C to higher than 250 °C. Extensive design work carried out over the last several years has led to the evolution of a simple low-concentration version of the ICPC that provides an elegant solution to several potentially expensive or difficult-to-implement features of previous ICPCs. This design was chosen to be used in a cooling demonstration project in Sacramento, California (Winston, Duff, O'Gallagher, Henkel, Christiansen, and Bergquam, 1999; see also Duff, Winston, O'Gallagher, Henkel, Mushaweck, Christiansen, and Bergquam, 1999).

The idealized optical design of this new configuration corresponds to the unit concentration limit for a vertical fin CPC solution, which is then coupled to a practical thin wedge-shaped absorber as shown in Figure 3.7. This design is extremely simple yet very effective. It has a concentric heat transfer tube that gives it rotational symmetry about the long axis so that operations on an automated glass lathe are greatly simplified. It maintains a relatively low fluid inventory per unit aperture by use of the ice cream cone-shaped absorber. Finally and most importantly, it doubles the effective concentration relative to the usual flat and horizontal fin absorber evacuated tube configuration (which loses heat from both sides). This, in combination with a good low emissivity-selective coating, is sufficient to reduce the thermal losses at operating temperatures between 150 °C and 250 °C to the levels associated with previous ICPCs. The absorber is coated with a newly developed proprietary high-temperature selective surface. The near-unit concentration ratio also allows a nearly full sky angular acceptance such that collection of diffuse radiation makes the thermal efficiency comparable to or better than a tracking parabolic trough at these temperatures. This design does not require a specially shaped concentrator profile to be incorporated by either a metal insert or reshaping of the glass tube. The concentrator is simply the silvered surface of the inside bottom-half circular cylinder of the glass tube. Thus, initial production and deployment of arrays capable of driving the new double-effect solar cooling systems can be accomplished at relatively low cost. The basic tube is 125 mm in outer diameter and 2.7 meters long. The absorber is a 0.2-mm-thick copper fin draped over and attached to a 12-mm copper heat transfer tube with a 7.5-mm-diameter thin-walled counterflow concentric feeder tube inside as shown in Figure 3.8.

Low Concentration Integrated Compound
Parabolic Concentrator (ICPC)

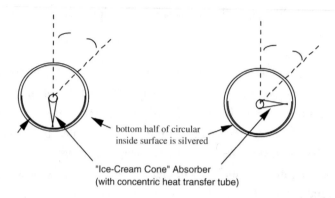

bottom half of circular
inside surface is silvered

"Ice-Cream Cone" Absorber
(with concentric heat transfer tube)

a) symmetric configuration          b) asymmetric configuration
(vertical fin)                (horizontal fin)

Unit Concentration Integrated CPC Tube
Concentric half-cylindrical configuration

Schematic Illustration

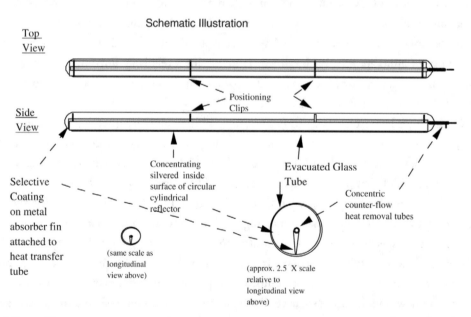

**FIGURE 3.7:** The limiting case for a vertical fin type absorber corresponds to a simple semicircular reflector. If the bottom half of the glass envelope is silvered, an effective concentration near unity (here, $C = 0.94$) is obtained, eliminating the need for an internal shaped reflector or shaping of the glass envelope.

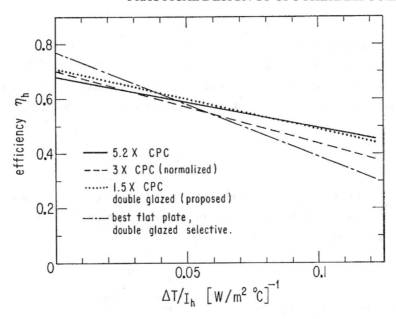

**FIGURE 3.8:** Thermal performance of nonevacuated CPCs developed in the early days of nonimaging optics (source: Rabl et al., 1980).

## 3.2   CPCs WITH NONEVACUATED ABSORBERS

When the thermal absorber is not surrounded by vacuum, the additional heat transfer from the hot absorber to ambient surroundings by both convection and conduction increases the thermal losses significantly so that the advantage of relatively low levels of concentration achievable with a CPC cannot be as dramatic as in the evacuated case. In principle, the losses are still proportional to the area of the thermal receive surface and so should scale approximately inversely with the geometric concentration ratio. However, they are larger, such that efficient delivery of heat at higher temperatures requires concentrations that cannot be achieved with a stationary collector. A fully stationary CPC with a nonevacuated absorber can still deliver respectable thermal performance, equivalent to that of a good flat plate. In addition, there can be cost advantages to replacing expensive metal absorber plates and back insulation with less expensive concentrating reflectors; we thus summarize here what has been learned about the design and performance of nonevacuated CPC thermal collectors. Comprehensive treatment of convective and conductive heat losses is discussed in the work of Rabl (1976b).

There was considerable effort devoted to developing nonevacuated CPCs in the mid 1970s, and many of the details of these studies were reported by Rabl, O'Gallagher, and Winston (1980).

### 3.2.1   Performance Model for Nonevacuated CPCs

The expressions defining the solar-to-thermal conversion efficiency for a nonevacuated CPC are very similar to those already discussed earlier (Eqs. (3.1)–(3.6)) for evacuated CPC collectors except that thermal loss expressions are not dominated by radiative terms. The optical efficiency is essentially identical to that given in Eq. (3.4) except that there is usually only one glazing with transmittance $\tau$ so that we can write

$$\eta_o = \tau \rho^{\langle n \rangle} \alpha (1 - L) \Gamma \tag{3.7}$$

In addition, it is more appropriate to combine all the heat loss terms into a single linear term with a heat loss coefficient, $U_L$, similar to that used for flat plate collectors so that the full expression for the thermal efficiency becomes

$$\eta(T_{fl}, T_{amb}) = \eta_o - U_L (T_{fl} - T_{amb}) \tag{3.8}$$

where $T_{fl}$ is the average fluid temperature usually approximated by

$$T_{fl} = (T_{out} + T_{in})/2 \tag{3.9}$$

and $T_{out}$ and $T_{in}$ are the outlet and inlet temperatures of the collector, respectively.

As examples of what can be accomplished with a seasonally adjusted nonevacuated CPC, we consider the two nonevacuated CPC prototypes that were designed and tested at the University of Chicago in the 1970s. These were described in detail by Rabl et al. (1980). One design utilized a vertical fin absorber and had a geometric concentration of 3×, whereas the other had a circular tubular absorber that was sized so as to correspond to a concentration of 5.2×. The 3× design had an acceptance angle of 18° and requires two adjustments per year (a winter position and a summer position), whereas the 5.2× design was designed for an acceptance angle of 6.5° and requires 12 adjustments per year. The thermal performance achieved for each of these is shown in Figure 3.8 and corresponds to heat loss coefficients of 2.7 and 1.9 W/(m²-C) for the 3× design and the 5.2× design, respectively. These loss coefficients are quite respectable when compared with values for typical flat plate collectors of 4.5–5.5 W/(m²-C). Note that, as expected, the heat loss coefficients scale roughly as $1/C$. As with evacuated CPCs, these higher-concentration devices requiring only seasonal adjustment may yet play a very significant role in making solar thermal energy a viable alternative at temperatures higher than 100°C, particularly in developing economies.

Before leaving the topic of nonevacuated CPCs, it is worth noting that a completely stationary design that employs additional heat suppression techniques may yet provide a stationary version

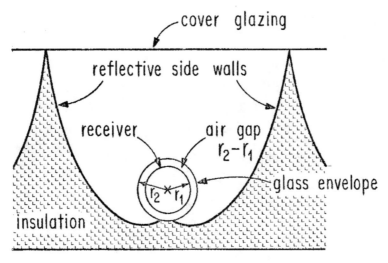

**FIGURE 3.9:** A "double-glazed" nonevacuated absorber can reduce thermal losses so as to provide a path to a high-temperature collector with a low-concentration nontracking CPC.

of this collector with exceptional thermal performance. One concept that has received some attention is that which surrounds a circular absorber with an additional glazing as shown in Figure 3.9. The optical losses introduced by this additional layer of glass somewhat depress the low-temperature performance, but these are more than offset by lower heat losses at higher temperatures. The predicted heat loss coefficient for such a collector is $U_L = 2.2 \, \text{W/(m}^2\text{–C)}$, corresponding to the straight-line performance for the 1.5× double-glazed CPC shown in Figure 3.8.

CHAPTER 4

# Practical Design of CPC PV Concentrators

## 4.1 ECONOMIC CONSIDERATIONS FOR PV APPLICATIONS

Although CPCs have been considered for use in combination with solar cells since the mid 1970s, PV application has remained much less developed than thermal application. There has never been a commercial CPC PV module developed to the point of marketability, although experimental prototypes and several alternative concepts have been fabricated and tested over the years. The reason for this lack of development is most probably that if the design is not carefully done, the complications introduced by the incidence of concentrated sunlight on the cell or cell array tend to offset the economic advantages of the relatively low levels of concentration achieved by single-stage CPCs. Nevertheless, the incorporation of CPCs into PV systems remains an appealing concept, so we review the basic tradeoffs and practical design considerations here.

Recall that the primary objective in employing concentration in a PV system is to reduce the cost per unit collecting aperture of the expensive solar cell by increasing the effective collecting aperture for a given area of solar cell. To understand the tradeoffs, we consider an admittedly oversimplified cost model. If the cost per meter squared for solar cells is $X$ and the cost per meter squared of concentrator aperture (reflector or lens material and support structure) is $Y$, where $Y$ is (presumably) much less than $X$, then $K$, the resulting net cost per unit aperture, is given by

$$K = Y + X/C \qquad (4.1)$$

where $C$ is the geometric concentration of the CPC. One replaces the expensive cell area by an inexpensive concentrator. In the limit that $C$ is very large, the cost of the cell becomes negligible. However, we immediately become aware of one limitation of a simple CPC. As we noted in Chapter 2, $C$ will be limited to about ≤10 for nontracking (seasonally adjusting) deployment geometries. A second limitation is encountered when one begins to consider the effect of even these low levels of concentration on cell array performance, which we do in the following section.

Before considering the direct effects of concentration on performance, it is useful to extend our simple economic model to include explicitly the cell and concentrator efficiencies, $\eta_{el}$ and $\eta_c$, respectively, and to relate all the system parameters to a simple economic figure of merit, such as the base module price $P$, in dollars per peak watt. We define $\eta_{el}$, the cell electrical efficiency, to be the ratio of the electrical power delivered by a cell (or cell array) to the radiant solar power (Insolation × Area) incident on the cell. Similarly, we define $\eta_c$ to be the ratio of the total radiant solar power delivered by the concentrator to the cell divided by the solar power incident on the concentrator aperture. For simplicity, we consider both $\eta_{el}$ and $\eta_c$ to be constants, although in practice they will vary somewhat depending on conditions. With these definitions, we see that we can express the base price for the cell (or array) by itself without a concentration as

$$P_o = X/(I_o \eta_{el}) \qquad (4.2)$$

where $I_o$ is some standardized reference value of the insolation (usually taken as 1000 W/m²). *If we assume that the PV efficiency remains the same with a concentrator,* we find that our economic figure of merit becomes

$$P_c = (Y + X/C)/(I\eta_{el}\eta_c) \qquad (4.3)$$

Finally, we see that, under these assumptions, the cost reduction factor $P_c/P_o$, achieved by incorporating concentration into our system is

$$P_c/P_o = (1/C + Y/X)/\eta_c \qquad (4.4)$$

For typical values of $C \cong 5$ , $Y/X \cong 1/5$, and $\eta_c \cong 0.8$, we see that the approximate cost reduction could be around a factor of 2 (or maybe 3) for a moderate concentration CPC system, *if the efficiency of the PV conversion remains the same.* Clearly, the cost of PV systems is in the range where this factor of 2 or 3 could be decisive in achieving economic viability. However, it is when the effects of concentration on cell performance are taken into effect that the difficulties in achieving such a breakthrough become apparent. It should be emphasized, however, that with careful design, they should be able to be overcome.

## 4.2    PERFORMANCE EFFECTS IN CPCs FOR PV APPLICATIONS

To illustrate the performance aspects, we consider the system shown in profile in Figure 4.1. Here, a flat absorber CPC is designed to be coupled to a single PV cell. This particular CPC was designed for a possible space application with a small acceptance angle (±5°) and truncated to the desired

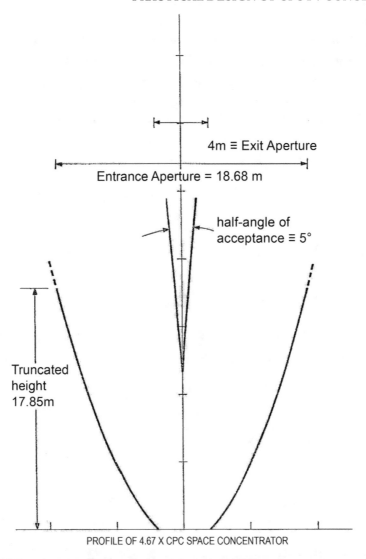

4m ≡ Exit Aperture

Entrance Aperture = 18.68 m

half-angle of
acceptance ≡ 5°

Truncated
height
17.85m

PROFILE OF 4.67 X CPC SPACE CONCENTRATOR

**FIGURE 4.1:** Profile of an ~5× CPC considered for possible PV application in space but used here to illustrate several features of CPC PV concentrators common to all applications.

geometric concentration (4.67×), but the resulting performance measurements and calculations are applicable to virtually any CPC PV embodiment. All of the sunlight incident on the aperture (4.67× wider than the exit aperture) within the acceptance angle will be directed, after zero, one, or more reflections, to the flat exit aperture, here assumed to be completely filled with active solar cell.

To see the effect of introducing CPC-type concentration onto a single PV cell, we show in Figure 4.2 the *I–V* curves for a cell with (1) no concentration ($C = 1.0$), (2) with $C = 2.93\times$, and (3) with $C = 4.75\times$. The incident sunlight (insolation) was approximately the same during the times that the three curves were measured. It should be emphasized that the cells used here were not designed for high concentrations—that is, they were optimized for operation at solar flux levels of about "one sun" or 1000 W/m². Note that for each *I–V* curve, as the load resistance is increased and

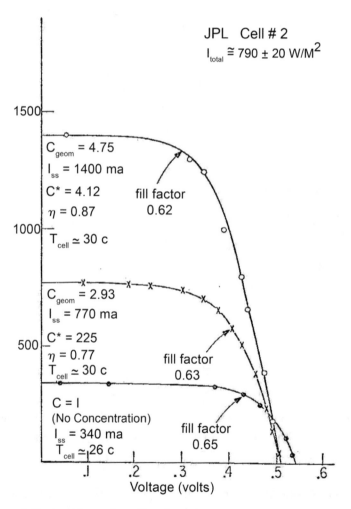

JPL  Cell # 2
$I_{total} \cong 790 \pm 20$ W/M²

$C_{geom}$ = 4.75
$I_{ss}$ = 1400 ma
$C^*$ = 4.12
$\eta$ = 0.87
$T_{cell} \approx 30$ c

fill factor
0.62

$C_{geom}$ = 2.93
$I_{ss}$ = 770 ma
$C^*$ = 225
$\eta$ = 0.77
$T_{cell} \approx 30$ c

fill factor
0.63

$C = I$
(No Concentration)
$I_{ss}$ = 340 ma
$T_{cell} \approx 26$ c

fill factor
0.65

Voltage (volts)

**FIGURE 4.2:** Three *I–V* curves for a solar cell under unit concentration, a geometric concentration ~3×, and a geometric concentration ~5×. The short-circuit current is a good measure of the optical efficiency of the concentrator, but the maximum power point also shifts under concentration such that the overall efficiency of the electrical performance needs to be carefully understood.

the output voltage and power increase, the current drops at first slowly and then more rapidly until the *I–V* product is a maximum at the maximum power point near the "knee" of the curve.

The best measure of the optical performance characteristics of a PV concentrator system is the short-circuit current of the PV cell. To a very good approximation, the short-circuit current is a direct measure of the net radiant energy falling on the cell. Based on this, the net "optical efficiency" values for the 2.93× and 4.67× concentrators are measured to be 0.77 and 0.87, respectively. (The difference is probably attributable to the use of different reflector materials: aluminized mylar and silver foil, respectively.) The short-circuit current measured as a function of the angle of incidence of the sunlight is used to measure the angular acceptance properties of the CPC. As shown in Figure 4.3, the response is roughly flat across the acceptance angle and then drops sharply at the acceptance angle but remains considerable for a wide range of angles due to the large partial acceptance outside the design acceptance associated with the rather severe truncation.

A good measure of the cell electrical performance is the "fill factor," defined as the ratio of the maximum power output to the product of the open-circuit voltage and the short-circuit current. The fill factor is indicated for each of the three curves in the figure. Note that this fill factor drops slightly as the concentration is increased. This is due to the increased current being drawn from the cell along with the internal series resistance of the cell. Recall that the cells were not optimized for

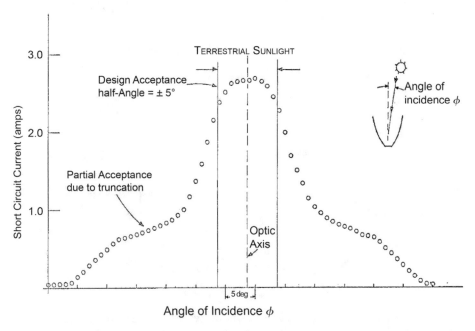

**FIGURE 4.3:** The angular acceptance function of an ~5× CPC coupled to a single PV cell. The short-circuit current provides the best measure of the optical performance of the concentrator.

performance under concentration. This would be a problem even if the intensity distribution was uniform across the width of the cell under concentrated sunlight, but this is hardly the case. The distribution of intensity across the cell at different angles of incidence as determined by ray trace is shown in Figure 4.4. At normal incidence ($\theta = 0°$), the effect of the two symmetric parabolic segments making up the CPC is to partially focus the extra reflected light into two symmetric "shoulders" in the intensity distribution. As the sunlight direction is moved off normal incidence, one of these shoulders grows while the other shrinks. Eventually, when the incidence angle is close to the

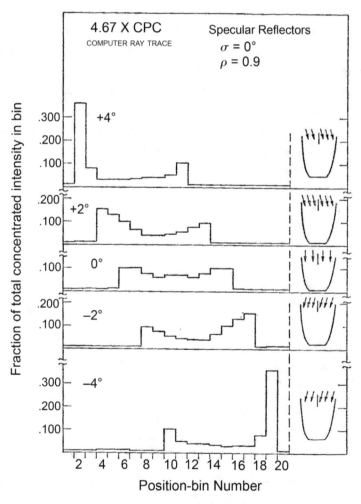

**FIGURE 4.4:** The intensity distribution across the transverse dimension of an ~5x trough is significantly nonuniform. This can have deleterious effects if multicell arrays are deployed across the target plane.

design acceptance angle, there is one very large peak of concentrated energy near the opposite side of the cell from the CPC sidewall where much of the light is being reflected.

If the PV device across the exit aperture of the CPC is a single crystal solar cell, the effect of this nonuniform intensity distribution across the cell is not too severe, although the high local current densities contribute to a slight drop in performance. This can be somewhat alleviated by using solar cells that are optimized for use under concentration. However, such cells are usually considerably more expensive than one-sun cells, often by factors of 5–10 or even higher on a cell area basis. Thus, the whole purpose of using the moderate levels of concentration delivered by CPCs will be defeated! The solution is to be very careful to keep concentration levels and nonuniform intensity levels in such a range that the one-sun cell performance is only marginally degraded. One approach that was investigated very early (Greenman, O'Gallagher, Winston, and Costogue, 1979) is to try to reduce the extreme intensity variations by introducing a small nonspecularity in the reflector surface. Alternatively, one could try to develop concentrator cells whose costs remain close to those of one-sun cells. To date, this has not been accomplished in a commercially viable system.

A further complication that also has a possibly deleterious effect on the system performance is the higher operating temperature of the PV cell that results from the incidence of concentrated sunlight even a few times higher than normal one-sun solar intensity levels. If one has to design into the system passive (or even active) means of cooling the cell, this can further increase the costs of the module and offset the desired economic benefits of concentration.

One more feature of CPC PV systems that needs to be mentioned is the very deleterious effect of nonuniform intensity variations if multiple-cell modules are used as the PV transducer at the CPC exit plane, rather than the single-cell configurations discussed up to now. Such arrays use series-parallel combinations to achieve the most desirable module voltage and current characteristics, which is fine when the intensity is uniform across the entire array. However, it must be recognized that the current in a series string of cells is *limited to the current in the least illuminated cell*. Thus, since the severe intensity variations take place across the width of the exit aperture, all the cells across the width of a target array must be connected in parallel with one another. Since the intensity along the direction of the trough (perpendicular to the plane of the CPC profile) is close to uniform, cells along the trough axis can be connected in series. However, one must be careful of end effects in such configurations.

It is probably the combination of the effects of the complications discussed in the proceeding paragraphs that has prevented the development of a commercial single-stage CPC PV system. Most of the prototypes developed followed the approach of coupling each individual cell to its own CPC trough. Since single-crystal cells are limited in width, this means that CPC modules of significant area are composed of a large number of individual troughs. This increases the complexity and costs of the support structure and concentrator configuration. As larger-width solar cells become available, this approach deserves to be revisited.

The other approach to overcoming these complications is to employ much higher levels of concentration so that the cost benefits apparent from Eq. (4.4) can be realized. This can be done even for quite expensive concentrator cells if the costs of the concentrator can be kept within appropriate limits. We shall discuss these matters further in Chapter 6.

. . . .

CHAPTER 5

# Two-Stage Nonimaging Concentrators for Solar Thermal Applications

## 5.1 BASIC CONCEPTS

So far in this monograph, we have limited our consideration of nonimaging solar concentrators to relatively low-concentration applications in which the optics employs only one stage of concentration. For many applications in solar energy, much higher geometric concentration values are necessary to achieve the desired performance or economic gains. Concentration factors of tens, hundreds, and even thousands are necessary. Obvious examples of high-concentration applications include solar thermal electricity generation, in which the temperatures required for efficient engines can only be achieved with very high concentrations, and PV applications, in which highly efficient but very expensive solar cells require that the cell area be kept as small as possible. For most thermal electric applications, relatively high temperatures (in the range of 500–1200 °C or even higher) are required. As one tries to operate at increasingly higher temperatures, radiation losses increase very rapidly and can only be overcome by applying increasingly higher geometric concentrations. However, one cannot continue to increase concentrations indefinitely.

Recall that by *geometric concentration* we mean here the ratio of total collecting aperture area to the final target area (as defined in Eq. (1.1)). In principle, by applying a concentration high enough, temperatures approaching the temperature of the surface of the sun (5800 °C) could be achieved. This would require a concentration given by Eq. (1.2b) with $n = 1.0$ in the limit that $\theta_c \geq \theta_s$, the half-angular subtense of the sun or 4.6 millirad. The corresponding geometric concentration (a factor of about 46,000) represents the maximum concentration achievable (in air or vacuum) as determined by the second law of thermodynamics and is referred to as the "thermodynamic limit." This limit is the maximum possible concentration allowed by physical conservation laws and is impossible to achieve with a focusing primary by itself. Any higher concentration would allow temperatures in excess of the solar source temperature to be generated, thus violating the second law. Therefore, attempts to develop a concentration higher than this thermodynamic limit (e.g., by reducing the size of the target still further) invariably must introduce throughput losses that rapidly reduce the net efficiency.

In principle, of course, single-stage CPCs could be designed with very small acceptance angles and resulting high geometric concentrations. However, for $\theta_c$ values of approximately 5° or lower, such devices become very tall and skinny and are too cumbersome to be practical. Therefore, to incorporate the benefits of nonimaging techniques in systems of higher concentration, one must resort to multistage configurations. It turns out that a variety of nonimaging optical devices, including the CPC, can be deployed in the focal zone of image-forming primary concentrators to increase the combined concentration substantially to higher than that achievable by the primary alone (Rabl and Winston, 1976; Welford and Winston, 1980). Such two-stage configurations have the capability of approaching the thermodynamic limit on a geometric concentration for a given angular field of view. In a point focus reflecting geometry, this typically means that the nonimaging secondary can deliver an additional factor of 2–4 in a concentration higher than that possible with the primary alone. In general, this is accomplished without doing anything to the primary. Nonimaging secondaries thus offer the solar concentrator designer an additional degree of freedom unavailable with any conventional approach. This unique capability can be used to increase either (1) the total concentration or (2) the effective angular field of view of the concentrating optics (or perhaps some of each). For solar thermal applications, these parameters are respectively related to increased performance (thermal efficiency) and relaxed optical tolerances (mirror slope error, angular tracking, etc.). These matters are discussed extensively in the work of O'Gallagher and Winston (1987a, b, c, 1988a, b).

A brief review of the development of two-stage nonimaging concentrators for high-temperature thermal and thermal electric applications and a preliminary comparison of the behavior of different types of secondary concentrators are presented in this chapter. Higher-concentration two-stage configurations for PV applications are addressed in the next chapter.

Let us review the design principles for optimized two-stage concentrators and discuss the characteristics and properties of each of the three types of nonimaging secondaries that have been employed. These are (1) CPC types and compound elliptical concentrators (CECs) as secondaries (Welford and Winston, 1980), (2) the so-called flowline or "trumpet" concentrators (O'Gallagher, Welford, and Winston, 1987) described in more detail subsequently, and (3) an alternative type of nonimaging concentrator referred to as a tailored edge-ray concentrator or TERC (Winston and Ries, 1993; Ries and Winston, 1994) that has been proposed for use as a secondary. Also discussed extensively in this chapter are the experimental development of secondary concentrators and some practical problems encountered in their use.

## 5.2   GEOMETRIC CONSIDERATIONS

The basic geometry for a two-stage system employing a nonimaging secondary is illustrated in Figure 5.1. The maximum relative benefit from a secondary will be provided in cases where $\phi$, the rim

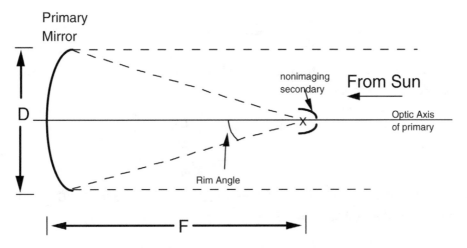

**FIGURE 5.1:** Schematic summary of the configuration for a two-stage nonimaging concentrator.

angle of the primary, is small, for systems with relatively large focal length-to-diameter (*F/D*) ratios, that is, for systems with relatively large focal length to diameter (F/D) ratios.

It is easily shown (Rabl, 1976a; Welford and Winston, 1978) that the concentration of a single-stage paraboloid or any other conventional reflecting focusing primary falls well short of the limits of Eq. (1.2b). In particular, for an incident light distribution characterized by a half-angle $\theta_i$ (defined to include the effects of sun size, slope and specularity errors, etc.) a primary with a convergence half-angle (or rim angle), $\phi$, as schematically illustrated in Figure 5.1, can attain a geometric concentration ratio without intercept losses, of at most

$$C_{1,\max} = \frac{\cos^2\phi\sin^2\phi}{\sin^2(\theta_i)} \qquad (5.1)$$

and thus falls short of the maximum concentration limit by a factor of $\dfrac{1}{\cos^2\phi\sin^2\phi}$. The highest concentration for a single stage given by Eq. (5.1) occurs for $\phi = 45°$, where $\cos^2\phi\sin^2\phi = 1/4$. Thus, even for this best single-stage concentrator, the shortfall with respect to the limit of Eq. (1.2b) is at least a factor of 4. However, by a suitable choice of secondary, one can recover much of this loss in concentration.

Nonimaging secondary designs are quite similar in concept to the single-stage CPCs already discussed in that their shape is developed by application of the so-called edge-ray principle (Welford and Winston, 1978, 1989) in the particular geometry being employed. For most types of conventional nonimaging secondary, the device is characterized by a design acceptance angle, $\phi_a$, and in three dimensions the corresponding secondary can achieve a geometric concentration factor of

$$C_{2,\max} = \frac{1}{\sin^2 \phi_a} \tag{5.2}$$

To accept all radiation from the primary, $\phi_a$ should be greater than or equal to the rim angle $\phi$, so that the practical geometric limit is

$$C_{2\text{-stage},\max} = C_1 \cdot C_2$$

$$= \frac{\cos^2 \phi}{\sin^2 \theta_i} \tag{5.3}$$

which comes close to the limit of Eq. (1.2b) for small $\phi$ (i.e., for large focal $F/D$ ratios). The rim angle $\phi$ is related to the focal ratio $f = F/D$ by

$$\phi = \tan^{-1}\left[\frac{1}{(2f - 1/(8f))}\right] \tag{5.4}$$

The combined effect of these relationships (Eqs. (5.3) and (5.4)) is summarized in Figure 5.2. Here, the maximum geometric concentrations for one- and two-stage concentrators, normalized to

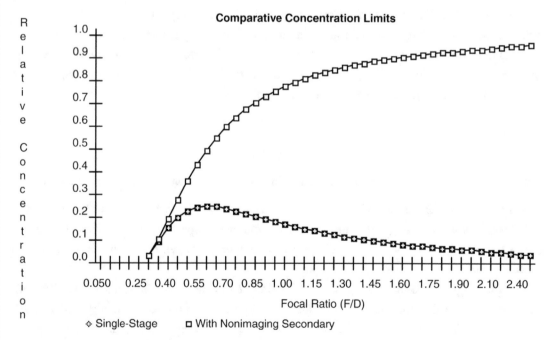

**FIGURE 5.2:** Comparison of the maximum geometric concentrations attainable by one- and two-stage systems with the ideal limit.

the ideal (thermodynamic) limit, are compared as a function of the focal ratio of the primary. The highest concentration designs are those in which the primary has a large focal $F/D$ ratio ($F/D > 1$), in which case the secondary can achieve very high-concentration factors and the combined two-stage concentration readily exceeds 80% or 90% of the thermodynamic limit. This is in contrast to conventional systems, which achieve a maximum concentration of <25% of this limit at small focal ratios ($F/D = 0.4$–$0.7$) corresponding to $\phi$ that is near 45°.

Nonimaging designs with a large focal ratio offer the opportunity for the incorporation of a number of additional built-in advantages. For instance, the primary concentrator can be a simple spherical mirror, since in a long-focal-length design, the optical broadening due to spherical aberrations is negligible. Alternatively, these designs encourage the use of faceted or segmented primary mirrors or reflectors formed from pressure-stabilized metal or film membranes. The latter type tends to form a sphere to a very good approximation. Finally, we note that the off-axis aberrations are not as severe in the case of long-focal-length designs as they are in short-focal-length systems such that, with a secondary, even off-axis geometries can attain very respectable concentrations. The off-axis geometry is discussed more thoroughly in Chapter 9 in the context of the analysis of the limits to concentrations for central receivers.

## 5.3    CONCENTRATION, OPTICAL QUALITY, AND THERMAL PERFORMANCE OF TWO-STAGE DISH CONCENTRATORS

A performance model to evaluate various two-stage solar concentrator design configurations (O'Gallagher and Winston, 1987a, b, 1988b) was developed to explicitly evaluate the advantages of nonimaging two-stage concentrators. A computer ray trace optimization of one- and two-stage concentrator designs was carried out to model system performance as a function of mirror surface optical error, primary focal ratio, receiver design operating temperature, and sun off-track error.

The performance model compared the performance of solar thermal concentrators both with and without secondaries using a methodology patterned after an approach originally introduced by Jaffe (1982). However, we incorporated two very important changes.

First, we used Monte Carlo ray trace calculations to determine the focal plane distributions. Jaffe approximated these distributions by single-parameter Gaussian normal functions. In each of these cases, the scale of the radial distribution in the focal plane of the primary is characterized by the combined root-mean-square (rms) optical error distribution, $\sigma$, incorporating sun size, primary slope error, and nonspecularity, among others. These distributions in turn are used to evaluate the tradeoff between energy intercepted by apertures of different sizes (related to the geometric concentration ratio) and receiver heat losses. Our approach not only provides a better representation of the actual focal plane distributions but also allows treatment of non-Gaussian optical error distributions, deviations from axial symmetry (e.g., off-track bias), and nonparabo-loidal contours.

Second, we modeled the optical efficiency of the secondary to account for variations in its throughput as a function of the fraction of energy it intercepts. Jaffe simply assumed one constant value, typically between 0.9 and 0.95, independent of the relative size of the secondary compared with the scale of the focal plane distribution. This was an artificially severe representation, particularly for Gaussian distributions, since it always introduces an optical loss for the two-stage system, whether or not the secondary is actually doing anything.

The details of the performance model are described elsewhere (O'Gallagher and Winston, 1987a, b, 1988b). The results of this analysis are summarized subsequently.

The single most important result of these comparative optimizations is that, for literally any case of practical interest, the efficiency of the two-stage system is always greater than that of the single-stage system if all the other design parameters are the same. The performance enhancements with a secondary, which are evident in all the optimized efficiency curves presented, are a direct measure of the relative importance of the additional factor of 2–4 in concentration made possible by

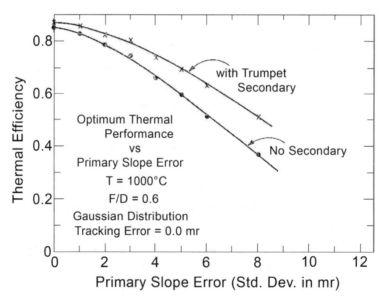

FIGURE 5.3: Illustration of the performance improvement provided by a trumpet secondary for the baseline case when the primary slope error is varied while the temperature is held constant. At small slope errors, the primary alone can achieve a concentration high enough to reduce heat losses to a negligible amount, such that the additional concentration with a secondary increases the efficiency only slightly. In this high-performance limit, a secondary can still be employed to increase tracking tolerances.

the secondary. The potential for improvement with a secondary can be understood very simply in terms of this relationship. If the marginal benefits of additional concentration are large, the performance gain is large; if these are small, the gain is small but usually significant.

### 5.3.1  Effect of Slope Error and Operating Temperature

Comparative optimized thermal conversion efficiencies determined by ray trace are shown as a function of primary Gaussian slope error at fixed temperature in Figure 5.3. As the mirror optical quality is improved, slope error is reduced and the optimized efficiency with and that without a secondary improve, with the two-stage system always remaining somewhat above the single-stage system, although the absolute and relative separations decrease as one approaches the limit of zero slope error. This is very similar to the behavior with respect to temperature at constant slope error as shown in Figure 5.4 and in both cases is a consequence of the diminishing importance of marginal increases in concentration when thermal losses are a relatively small fraction of the optical gain.

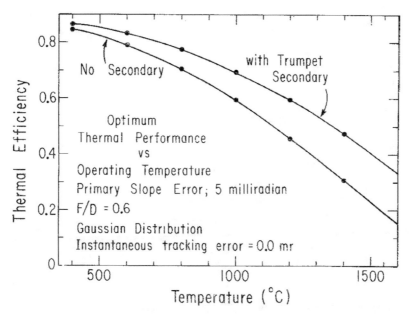

FIGURE 5.4: Illustration of the performance improvement provided by a nonimaging trumpet secondary as a function of design temperature. The primary is a reference baseline paraboloid (O'Gallagher and Winston, 1987a, 1987b) with relaxed optical errors corresponding to a characteristic slope error of 5 millirad. The solid points are optimized efficiencies calculated by balancing thermal losses against intercept gains using a focal plane distribution determined by Monte Carlo ray tracing.

### 5.3.2 Effect of Focal Ratio

There are two effects expected to be important as one varies the design focal ratio (*F/D*) of the primary:

1. Whereas the image size of an extended object produced by a parabola decreases with increasing rim angle, the off-axis optical aberrations (coma) increase. The combination of the two effects is a minimum at a rim angle of 45°, in that, for perfect optics, the corresponding "focal patch" size is smallest relative to the primary diameter. This means that, for a 100% intercept factor, the maximum geometric concentration occurs for this geometry, which corresponds very closely to *F/D* = 0.6. Thus, for a single-stage paraboloid with some distribution of optical errors, we expect performance to be optimum near this value of *F/D*.

2. If we add a secondary, the acceptance angle *C*, required for it to "see" the primary, decreases with a longer focal ratio, permitting higher secondary concentration ratios (Eqs. (2)–(5)). This effect more than offsets the decreasing concentration ratio of primary, such that the combined two-stage concentration asymptotically approaches the thermodynamic

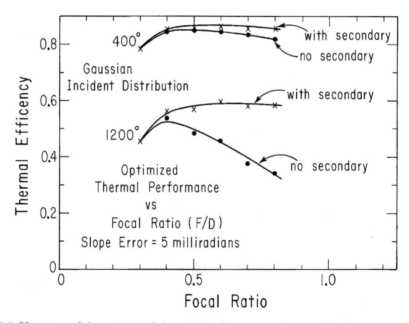

**FIGURE 5.5:** Variation of the optimized thermal performance at low and high temperatures with and that without a nonimaging secondary as a function of the primary focal ratio. At high temperatures and longer focal lengths, in which concentration to reduce thermal losses is necessary, the two-stage system maintains respectable performance, whereas loss of concentration of the primary alone causes its performance to fall off. The precise shape of these curves depends on the shape of the optical error distribution.

or "ideal" limit as $F/D$ increases. To the extent that the thermal efficiency depends on concentration, we should expect the performance of a two-stage system to exhibit a similar relationship relative to its focal ratio.

The combination of these effects can be seen in Figure 5.5 at two temperatures. The separation between the two curves is quite large at high temperature but small at low temperature in which the single-stage performance is only weakly dependent on the focal ratio. This is an indication of the relative lack of importance of increased concentration at low temperature.

### 5.3.3   Off-Track Tolerances With and Without Secondaries

As can be seen in Figure 5.3, the relative efficiency gain with a secondary is marginal in the limit of small slope errors and/or low receiver operating temperatures. This, of course, is because in this limit, one already has all of the concentration one needs to reduce heat losses to a negligibly small fraction of the optical gain with the primary alone and the additional factor of 2–3 from the secondary will not substantially improve the instantaneous performance. However, in this limit, there are

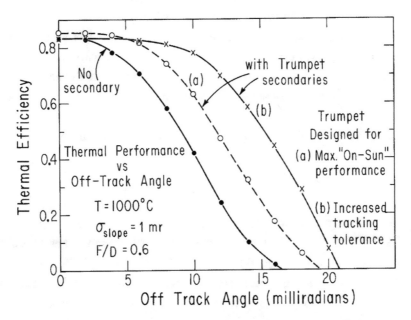

**FIGURE 5.6:** Thermal performance with and that without secondaries versus tracking bias error in the "high-performance limit" of a very small slope error. The secondary, which maximizes the two-stage on-sun efficiency, provides a small increase in angular acceptance as well, but a two-stage system designed to have the same on-sun efficiency as the optimized single-stage system can nearly double the allowed off-track error.

other optical benefits from a secondary that do not show up on efficiency graphs, such as Figures 5.3, 5.4, and 5.5. In particular, we refer to increased circumsolar collection and relaxed tolerance for off-track errors. The latter effect is shown in Figure 5.6 as an illustration of the magnitude of this type of benefit.

## 5.4 PERFORMANCE BENEFITS FOR SECONDARIES COMBINED WITH SPHERICAL MIRRORS

One can easily argue that the optimum application for secondaries lies in their combination with potentially inexpensive approximations to a paraboloidal figure, rather than with a true paraboloid. One example of such a configuration is a set of flat mirror tiles arranged appropriately to be tangential to an imaginary paraboloidal surface. Another possible example is a spheroidal mirror. Both of these cases (and perhaps many others) share some common features that make their combination with secondaries particularly attractive. First, they become better approximations to a paraboloid at longer focal ratios. Second, they appear to be less problematic to fabricate at longer focal ratios (fewer facets required in the first case, a shallower draw in the second). Since the secondary concentration ratio is increasing at longer focal ratios just slightly more than enough to compensate for the increasing image size due to the primary, it certainly seems like an ideal marriage! As a preliminary study of the types of performance that might be possible with such a compound design, the paraboloidal primary was replaced with a simple spherical mirror in the ray trace code and a set of optimization studies similar to those already discussed was carried out. The results can be summarized as follows.

When a concave spherical reflecting surface with a radius of curvature $K$ is used as a focusing mirror, its paraxial focal length is

$$F_o = K/2 \qquad (5.5)$$

However, for large rim angles or short effective focal lengths, rays reflected off the outer edge of the spherical surface will be focused at a point somewhat inside $F_o$, designated as $F_e$. This means that the focal plane distribution will change its shape between these two foci (15) and any optimization procedure that depends on this shape must take this variation into account. It turns out that the optimum location for a wide range of temperatures is about two-thirds of the distance from $F_o$ toward $F_e$. The relative performances of a spherical primary optimized with and that without a secondary were evaluated for values of $F_o/D$ between 0.6 and 1.2. The results are plotted in Figure 5.7 for the spherical analogue of our baseline case. Also shown for comparison is the performance of a single-stage true paraboloid. These results appear quite encouraging. The two-stage performance levels off at a value about 35% better than the best single-stage performance level. The latter level occurs at a

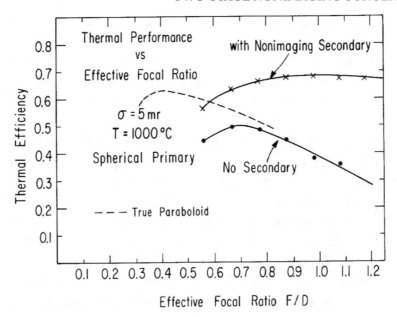

**FIGURE 5.7:** Thermal performance for the baseline optical errors and operating temperature, for single- and two-stage concentrators using a spheroidal primary, as a function of the actual focal ratio. The corresponding optimized performance for a single-stage paraboloidal contour is shown for reference. The two-stage spheroidal primary at moderate to long focal ratios is dramatically better than the spheroid alone and even exceeds the performance of the short-focal-length paraboloid. It is suggested that this may have possible applications with stretched membrane primaries not requiring preforming.

nominal focal ratio of 0.7, whereas the two-stage performance level is almost independent of $F/D$ for values $>0.8$.

Such designs or some variations on this theme may offer a very attractive path to economical delivery of solar thermal energy, particularly in view of developments in stretched membrane technology.

## 5.5   SUMMARY OF ADVANTAGES OF TWO-STAGE NONIMAGING CONFIGURATIONS

Based on the assumptions defined in the main body of this report and the exercising of the models developed, we have reached the following conclusions:

1. One will never lose performance by adding a secondary; one can only gain or break even.
2. In general, the optimized thermal efficiency for a two-stage system lies significantly above that for a single-stage system if all other characteristic parameters are the same for both systems.

3.  Based on investigation of the relative performances of one- and two-stage systems in the limit in which there may be only a small relative efficiency gain (e.g., low operating temperatures), we found that in such cases, one can design for a large gain in off-track tolerances, typically about a factor of 2, relative to those corresponding to single-stage systems.

4.  Based on investigation of the relative performances of the two types of systems when the primary is spheroidal rather than paraboloidal, we found that the optimum performance with a secondary is much better than the best performance with no secondary, and this optimum performance occurs at a focal ratio somewhat larger ($f = 0.8$–$1.0$) than that typical for conventional designs. This may have important implications for the use of primaries utilizing stretched membrane technologies.

5.  Using a methodology for the rational optimization of performance versus cost, we found that unless all costs are virtually independent of optical errors, a two-stage thermal system, so optimized, must always be cost-effective relative to the corresponding single-stage system (provided of course that the relative costs of the secondary itself remain small compared with the primary). The method is based on what should be a self-evident constraint—namely, that at the optimum, the relative incremental performance gains with respect to a particular performance parameter should balance the incremental costs associated with improvements in that parameter.

## 5.6    COMPARISON OF VARIOUS NONIMAGING SECONDARY CONCENTRATOR DESIGNS

### 5.6.1    CPCs and CECs

CPCs and CECs are two of the original family of nonimaging ideal concentrator solutions generated by application of the so-called edge-ray principle and were the first types to be considered as secondaries (Welford and Winston, 1980). They are characterized by relatively high reflection losses (the average number of reflections is slightly over 1.0) and skew-ray losses (about 5%), but they can achieve high geometric concentrations without large intercept losses in large focal ratio geometries (rim angles <15° or so). A CPC is designed applying the edge-ray principle for a fixed acceptance angle $\phi_a$, and the resulting CPC can usually be truncated to about half or even one-third of its fully developed height without significant loss in concentration. A CEC is designed by applying the edge-ray principle to the edge ray from the outermost piece of the reflector on the primary. This construction should be used in two-stage systems for which the net geometric concentration is low (100–500×). At high concentrations, the difference between a CPC and a CEC is not significant.

## 5.6.2 Flowline or Trumpet Concentrators

This type of concentrator is one of a second family of ideal nonimaging concentrators discovered (O'Gallagher et al., 1987). It has a number of advantages over CPCs in short focal ratio geometries that arise often in retrofit applications since single-stage concentrators are often optimized in this region (Figure 5.2). The trumpet's effective aperture is in the same plane as the physical exit aperture, which is another useful feature in retrofit applications. The trumpet basically is composed of a hyperboloid of revolution with an asymptotic angle that must be at least as large as the rim angle to intercept all the radiation from the primary. In practice, the trumpet is truncated by widening its asymptotes several degrees above the rim angle. It has relatively low reflection losses (typical average number of reflections is about 0.3) and no skew-ray loss at all (O'Gallagher, Winston, and Welford, 1987). It should be noted that, in the case of the trumpet, $\phi_a$ is the asymptotic angle of the hyperbola of revolution defining the flowline concentrator. In the limit that $\phi_a$ approaches $\phi$, the untruncated trumpet reaches the very edge of the primary, an obviously impossible solution to implement. When $\phi_a > \phi$ by even a small amount (2°–3°), the trumpet size becomes much smaller (O'Gallagher, Winston, Diver, and Mahoney, 1995, 1996). Further savings in size, cost, and shading caused by the trumpet can be achieved by truncating the reflector at the height at which the flux reflected by the primary is equal to that of the sunlight striking the back surface.

The generalized equation describing the profile for a trumpet secondary concentrator is

$$R(z) = \sqrt{R_x^2 + \frac{z^2}{(C_t - 1)}} \qquad (5.6)$$

where $R(z)$ is the radius of a hyperboloid of revolution at a height $z$ above its exit aperture; $R_x$ is the radius of the hyperboloid at its narrowest section, which is fixed as its exit aperture (corresponding to the physical receiver aperture); and $C_t$ is the geometric concentration of the trumpet. The trumpet acts as a concentrator because every ray entering the aperture of the "bell" of the trumpet, whose intersection with the target plane lies within a circle of radius $R_v > R_x$ (called the "virtual target"), must, after zero, one, or more reflections, pass through the actual exit aperture. Thus, the trumpet concentration ratio is given by

$$C_t = \frac{(R_v^2)}{(R_x^2)} \qquad (5.7)$$

which in turn is determined by $\psi$, the asymptotic angle of the hyperbola, such that

$$C_t = \frac{1}{\sin^2 \psi} \qquad (5.8)$$

The design parameters for any trumpet are then (1) $R_x$, which sets the actual scale of the concentrator in the focal plane; (2) $\psi$, which determines the concentration ratio and virtual target radius (i.e., the geometric optical intercept for a given focal plane distribution); and (3) $h_t$, the truncated height of the trumpet. Two trumpet profile solutions are illustrated in Figure 3 corresponding to $C_t = 1.70$ ($\psi = 50°$) and $C_t = 1.84$ ($\psi = 47.5°$). Full optimization of a trumpet design requires a series of ray traces comparing the performance tradeoffs while varying all three of these design parameters.

The truncation height $h_t$ and $\psi$ are independent parameters that affect the geometric throughput for a particular configuration, which in turn is determined by ray tracing over a range of possible choices. Clearly, $\psi$ should be $\geq \phi$, the rim angle of the primary. In the limit in which $\psi = \phi$, the trumpet would have to extend all the way to the primary, thus blocking it completely. As $\psi$ becomes just slightly greater than $\phi$, the asymptote of the trumpet intersects the extreme ray (extending from the edge of the primary to the edge of the virtual target circle) at a point much closer to the focal plane, such that physically compact trumpet concentrators can be readily generated. In fact, $h_t$ can be such that the physical bell of the trumpet need not extend all the way to the extreme ray, if one accepts the loss of a small number of extreme rays. Evaluation of the relative importance of any such

## Trumpet Profiles

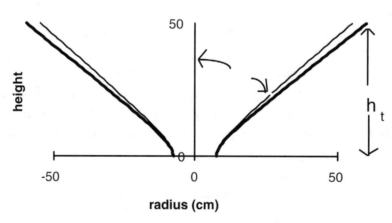

FIGURE 5.8: Illustration of two trumpet profile solutions. Each is a hyperboloid of revolution, here with a fixed exit radius of 7.62 cm. The geometric concentration ratio of each depends on the asymptotic angle of its hyperbola, and the throughput depends on the truncation height. The optimum design for the prototype test (see subsequent discussion) was selected after ray tracing a range of possible solutions.

losses, as well as studies of the actual reflection losses, requires a detailed ray trace of any particular design. These geometric losses are determined by detailed ray traces, and the actual trumpet selected is chosen so that they are negligible (the order of a few tenths of 1%).

## 5.6.3  TERCs

In the early 1990s, a new type of nonimaging secondary that in principle can achieve geometric concentrations close to 90% of the thermodynamic limit, even in short-focal-length geometries (i.e., focal $F/D$ ratios of 0.6–0.7), was developed (Winston and Ries, 1993; Ries and Winston, 1994). This could be very useful not only in retrofit applications but also in developing mechanically practical high-concentration two-stage systems. These systems are designed by reconfiguring the system slightly and taking advantage of the fact that not all of the aperture of the secondary is illuminated uniformly within its design acceptance angle (i.e., not all of the available phase space

| TABLE 5.1: Comparison of TPSC With Trumpet (CPG configuration: $F = 5.4$ m, $D = 9.6$ m, $\phi = 48°$) | | | |
|---|---|---|---|
| **PARAMETER** | **TRUMPET** | **TPSC 1** | **TPSC 2** |
| Height | 10.5 in (26.7 cm) | 10.5 in (26.7 cm) | 42.5 in (108 cm) |
| Entrance diameter | 25.7 in (65.0 cm) | 33.6 in (85.0 cm) | 67.0 in (170 cm) |
| Secondary concentration | 1.71× | 2.11× | 3.08× |
| Exit diameter | 6.0 in (15.2 cm) | 5.4 in (13.8 cm) | 4.4 in (11.4 cm) |
| Skew-ray loss (geometric; %) | 0 | ⪅1 | ⪅1 |
| Net reflection loss (reference = 0.8; %) | ~2 | ~5 | ~10 |
| Thermal load (kW) | 1 | ~2.5 | ~5 |

[a]Truncated to the same depth as trumpet.
[b]At maximum concentration for this rim angle.

is utilized). The secondary is then constructed using the actual extreme edge-ray incident at each individual point on its reflector surface. The more traditional "ideal" optical elements, such as the trumpet and CPC or CEC, are designed for one particular edge ray over the whole aperture and achieve secondary geometric concentrations given by Eq. (5.3). This yields system concentrations 50–65% of the full thermodynamic limit for the short focal ratio geometries referred to earlier. The actual concentrations attained by the new concentrator designs are somewhat dependent on their degree of truncation and the associated partial shading of the primary.

A test design case based on the geometry of an experimental test of a trumpet secondary is described in more detail in the next section (the CPG-460) to carry out a meaningful comparison. It was found that these *tailored phase space concentrators* (TPSCs, also sometimes called TERCS) can achieve geometric concentrations of 2.11× and 3.08× (1) when truncated to the same depth as the prototype trumpets shown in Figure 5.8 and (2) when concentration is maximized for this rim angle without regard for compactness, respectively. The results of this preliminary comparison are summarized in Table 5.1. The two TPSCs would have skew-ray losses of $\lesssim$ 1% (compared with zero for the trumpet) and reflection losses (assuming a secondary reflectance of 0.8) between 5% and 10% (compared with <2% for the trumpet). It should be noted that the TPSCs are often very well approximated by straight-sided cones and would appear to be particularly useful if high secondary reflectivities can be maintained.

## 5.7   SOME GENERAL OBSERVATIONS

In designing solar thermal systems with nonimaging secondaries, the main objective is to maximize the system concentration ratio to approach the thermodynamic limit as closely as possible. Thus, for conventional nonimaging secondaries (trumpets or CPCs) for which the combined geometric concentration is given by Eq. (5.4), one would like to make $\phi$ as small as practically possible—that is, to make $f$ as large as practically possible. However, very large focal length geometries are mechanically and operationally cumbersome. Furthermore, as shown in Figure 2, much of the improvement of the two-stage system is attained even at moderate focal ratios for which $0.7 \lesssim f \lesssim 1.0$ such that it is in this regimen in which most practical solar thermal two-stage concentrators will lie. In furnace applications in which very high concentrations are desired, longer-focal-length designs with $1.5 \lesssim f \lesssim 2.2$ have been found useful (O'Gallagher and Winston, 1988b). In principle, one could use either a CPC or a trumpet in any of these geometries; however, in practice, trumpets are more useful in short-focal-length applications ($f \lesssim 0.9$) with low secondary concentration, whereas CPCs are better in long-focal-length geometries ($f \gtrsim 0.9$) with high secondary concentration. For these cases, of course, $\phi_a$ for either type of secondary must be slightly larger than $\phi$ to accommodate extreme rays. One other practical limitation in all cases is the size of the secondary, such that usually the secondary is truncated somewhat relative to its fully developed height.

## 5.8   PRACTICAL CONSIDERATIONS AND AN OPERATIONAL TEST

As we have seen, the use of a secondary in combination with a focusing primary permits either the recovery of significant intercept losses while maintaining a fixed geometric concentration or substantial increases in geometric concentration while keeping intercept losses negligible. This can be done without requiring any improvement in the optical quality of the primary. The value inherent in providing these tradeoffs by using some type of secondary for increasing the thermal efficiency of concentrators used for solar thermal electric power has been the subject of considerable analytic work that had been summarized in the proceeding sections of this chapter. All these efforts have indicated that there is a strong potential for significant improvement in both optical and thermal performances at modest cost. Three experiments investigating the practical implementation of the secondary concept were carried out in their early development (Ortabasi, Gray, and O'Gallagher, 1984; O'Gallagher and Winston, 1986; O'Gallagher, Winston, Suresh, and Brown, 1987) to evaluate these concepts. All these were carried out with cold water calorimeter receivers. All showed significant gains in optical concentration and increases in geometric intercept, and all demonstrated the practicality of the concept as long as the secondary was actively cooled (by circulating cooling water through coils in the secondary). Some tests and supporting analytic work (Suresh, O'Gallagher, and Winston, 1987) showed that, under certain conditions, passive cooling techniques could be effective. However, they also revealed the vulnerability of the secondaries to problems of overheating if careful attention is not paid to the effectiveness of the cooling mechanisms.

An opportunity finally arose in the mid 1990s to carry out an experimental test of a trumpet-type nonimaging secondary concentrator (O'Gallagher and Winston, 1986; O'Gallagher et al., 1987) in combination with a prototype commercial dish–Stirling solar electric generating system. This was the Cummins Power Generation CPG-460 7.5-kW$_e$ concentrator system (Bean and Diver, 1992). The experiments were the first that had the capability of evaluating the performance of a secondary with a high-temperature receiver.

This experiment had the objective of evaluating the performance of a secondary operating in combination with an actual high-temperature receiver. The secondary was of the trumpet type, and the initial tests were done using the CPG-460 and a heat-pipe receiver equipped with a gas-gap calorimeter. The heat-pipe receiver and gas-gap calorimeter enable direct measurement of collector (concentrator and receiver) performance in which the receiver is maintained at a controlled high temperature, in this case, 675 °C.

The trumpet design selected was matched to the focal plane distribution of a new CPC-460 facetted membrane concentrator and corresponded to a hyperboloid of revolution with an asymptotic angle of 50° and a virtual target diameter of 19.9 cm (7.8 in) corresponding to a secondary geometric concentration ratio of 1.7×. The wider acceptance angle allows the trumpet to be truncated to a height of 27.4 cm (10.8 in) with negligible intercept loss. Ray trace analysis predicted an increase of about 3.5% in the total intercepted power, an increase in off-track tolerance at 90% intercept from 1 to 5 millirad, and a significant reduction in thermal losses from the smaller cavity aperture.

The basic objectives of these tests were twofold. First, we wanted simply to demonstrate the practicality of the use of secondaries of this type with a high-temperature thermal receiver. Second, we wanted to measure quantitatively the effect of the secondary on both optical and thermal performances of the receiver and to compare these results with model predictions. With regard to the first objective, we investigated at high temperature a number of operational concerns, including (1) the effectiveness of the active water cooling and (2) the effectiveness of the thermal isolation of the trumpet from the hot receiver.

Throughout the project, a conservative design philosophy was adopted. Thus, the baseline system was a retrofit design that did not attempt to optimize the performance of the full two-stage system. However, the experience we gained in this experiment provided much new practical information concerning this approach so that the full potential of the concept can be utilized in future applications.

We achieved all our qualitative objectives in a few days of on-sun testing with the trumpet in place. In particular:

1.  There was no fundamental operational problem in operating a water-cooled secondary in the immediate vicinity of a very hot (660 °C) cavity receiver. The trumpet throat temperature remained lower than 100 °C throughout the tests.

2.  There was no evidence of significant direct heat loss from the hot receiver to the cooled trumpet: That is, these tests showed that the thermal isolation of the trumpet from the hot receiver was very effective.

3.  Our experience on this and previous experiments repeatedly showed that the optical quality of any primary can be expected to fall well below design goals and deteriorate further with time. This expectation should be taken into account in planning experiments and developing concentrating systems. Due to this poor match between optical quality for which the trumpet was designed and the actual dish on which these first experiments have been carried out, the performance benefits associated with the trumpet were not accurately measurable from these tests.

## 5.8.1  The Importance of Maintaining the Optical Performance of the Primary

We have noted that the optical quality of the facets on this dish is known to have undergone some deterioration due to stretching and/or sagging after prolonged exposure to moisture. At the time the secondary was originally designed, in the summer of 1993, it was found that the measured focal plane distribution for newly fabricated and aligned facets could be well approximated by a circular Gaussian of the form

$$P(r) = P_o \exp \left[ -\frac{r^2}{r_o^2} \right] \qquad (5.9)$$

Here, $P_o$ is the peak power per unit area at the center of the distribution and $P(r)$ is the power per unit area at a radial distance $r$ away from the center (Jaffe, 1982) and $r_o$ is the characteristic rms radius of a given actual distribution. For purposes of designing the secondary, such primary focal plane distributions can be simulated by a comprehensive Monte Carlo ray trace model whose parameters can be varied to obtain the best fit to the observed radial distribution. The initial optical quality of the Cummins primary was well characterized by slope and specularity errors of 2.1 and 1.5 millirad, respectively, and a Gaussian sun of rms angular subtense 2.73 millirad. This yielded a total effective rms angular spread of $\sigma = 5.2$ millirad and corresponded to a radial scale of $r_o = 3.5$ cm. This distribution was an excellent match to the initial focal plane distribution measured in early 1993 for the new, well-aligned facets. The optical design for the trumpet used in these experiments was based on these parameters and made no allowance for the subsequent deterioration.

However, in the actual experiments, it was found that the thermal load on the trumpet (measured from the temperature rise in the cooling water) was very much larger that expected. Thus, it appears that the characteristic scale of the focal plane distribution increased significantly due to the deterioration of the optical quality of the facets between the time of the original focal plane mapping (1993) and the time of these first detailed trumpet measurements in December 1995. For a given characteristic radial scale $r_o$, one can calculate the intercept factor $\Gamma$, for a given aperture radius $R$. In particular, it is easy to show that

$$\Gamma = 1 - \exp \left[ -\frac{R^2}{2r_o^2} \right] \qquad (5.10)$$

An aperture diameter of 7.0 in (17.8 cm) corresponds to a radius $R = 8.9$ cm, and the upper limit value of $\Gamma = 0.69$ calculated previously can be used in Eq. (5.10) to solve for a lower limit on the actual characteristic scale. Such a calculation yields $r_o = 5.81$ cm. This effect of the primary deterioration can be appreciated by comparison of this scale with the original value observed in 1993 of $r_o = 3.5$ cm for which the trumpet was designed. The difference is due to sagging and stretching of the stretched membrane facets during the intervening 2 years. These experiments have made a good beginning on understanding that much remains to be done to achieve our quantitative goals. However, the experience we have has generated further confidence in the approach.

CHAPTER 6

# Two-Stage Nonimaging Concentrators for Solar PV Applications

We have seen in Chapter 4 how PV cells designed for one-sun (no concentration) operations have usually been found to be unsuitable even for relatively low levels of concentration. Practical problems associated with accommodating nonuniformities and higher operating temperatures usually result in some need to use more expensive cells, thus offsetting the economic benefits of concentration. In fact, the much higher costs associated with the need to use high-concentration cells drive the levels of geometric concentration required to realize a meaningful economic benefit up by a couple of orders of magnitude to an approximate $C_{geom} \geq 100\times$, well beyond the range attainable by single-stage CPCs. It is beneficial to use a two-stage configuration to attain these levels.

As we found in Chapter 5 for high-temperature solar thermal applications, it turns out that such a two-stage concentrator configuration combining a focusing primary with a nonimaging secondary concentrator can achieve the same fundamental advantages that we have come to associate with nonimaging optics—that is, either substantially higher concentration for a given set of optical tolerances or greatly relaxed optical tolerances for a given level of desired concentration. However, most of the two-stage PV concentrators will look much different from the two-stage solar thermal concentrators described in Chapter 5.

## 6.1 MULTIELEMENT CONCENTRATOR ARRAYS

The usual pathway for the PV conversion of solar energy to electricity using concentration is to use individual cells operated under concentrated solar flux. However, this approach imposes the additional requirement that each cell has to have its own concentrator. This is because the current in a series string of cells is limited to that generated in the cell with the lowest irradiance. Thus, in general, if a series-parallel array is operated with a single large concentrator, the resulting nonuniform irradiance on the array degrades its performance to the point at which such designs are unworkable. Thus, virtually all concentrating PV systems developed up to now employ modular arrays of small concentrators combined with individual cells. These systems require that the optical axes of the array of concentrators be mutually aligned with one another to tight tolerances, and the whole array is then tracked together to follow the sun. Even in this case, the nonuniformity on a single cell can degrade the performance.

The scale of each single element in the array is thus determined by the size of the individual concentrating PV cell. We consider the design of these individual elements first. By and large, the most common primary PV concentrator is some kind of lens, since any structure required for cooling the cells under concentration and supporting the electrical interconnections between cells would shade a reflecting primary. Since the cells are usually limited to the order of 1 to a few centimeters in diameter, the concentrating elements with geometric concentrations of one to a few hundreds have collecting aperture diameters in the order of tens of centimeters. This assumes that a circular cell will have a circular aperture. Of course, the individual elements constituting an array are usually close packed and the apertures are often rectangular producing a mismatch in symmetry between the aperture and the cell.

## 6.1.1   Geometric Considerations

The typical geometry of a two-stage nonimaging PV concentrator is shown in Figure 6.1. The nonimaging concentrator is of the dielectric CPC (DCPC) type already described in Chapter 2 with an acceptance angle designed to "see" the largest transverse dimension of the primary (e.g., the diameter of a circular lens or the diagonal of a square lens). The secondary aperture placed roughly in the focal plane of the primary lens enlarges the effective size of the PV cell so that the optical quality of the primary/and or the required tracking tolerances can be relaxed considerably with respect to what they would have to be if the cell alone was placed in the focal plane. Alternatively, the relative size of a large (expensive) concentrator cell/primary-only lens configuration, with a given optical quality, can be greatly reduced by employing the secondary. An additional benefit is that the secondary also serves to reduce the nonuniformities in the concentrated solar flux incident on the cell. A comprehensive treatment of the design of these two-stage PV concentrators and appropriate secondary configurations is given by Ning, O'Gallagher, and Winston (1987a, b).

The nonimaging secondary is formed from a transparent dielectric material with an index of refraction $n = 1.3–1.5$. It is usually possible to ensure that total internal reflection occurs for all rays accepted by the primary (Winston, 1976). Thus, the costs associated with applying and protecting metallic reflecting surfaces can be avoided. Furthermore, the refractive power of the dielectric operates in combination with the reflective sidewall shapes such that the secondary concentration ratios achievable are a factor of $n^2$ larger than those possible using conventional reflecting secondaries. The primary is a lens (usually a Fresnel lens, since this can be thin and lightweight) of aperture diameter $D$. The primary lens focuses normally incident rays at a point that defines the center of the primary receiver. Rays from the outer edges of the primary are brought together with a convergence angle

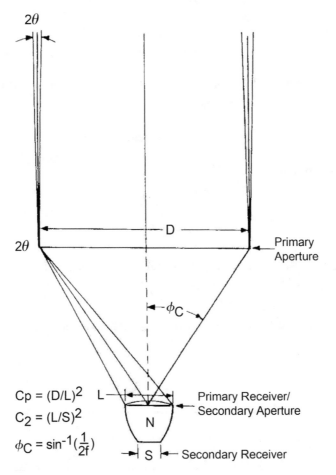

**FIGURE 6.1:** Schematic diagram of a two-stage point focus PV concentrator. The primary would typically be a Fresnel lens. The secondary is a DCPC-type element with an index of refraction $n$.

$2\phi_c$. If the incident rays subtend an angle of $\pm\theta_p$ on either side of the aperture normal, the focal spot will extend over a region here designated by diameter $L$ such that the primary concentration ratio $C_p$ is

$$C_p = D^2/L^2 \qquad\qquad (6.1)$$

The maximum possible geometric concentration $C_{max}$ for the system from Eq. (1.2b) in a point focus geometry is

$$C_{max} = n^2/\sin^2\theta_p \qquad\qquad (6.2)$$

for a focusing lens of focal ratio $f$ where

$$f \equiv 1/2\sin\phi_c \qquad (6.3)$$

corresponding to the generalized focal ratio used to express the Abbe sine condition for off-axis imaging (Welford and Winston, 1978) where

$$\pm L/2 = \pm D \cdot f\sin\theta_p \qquad (6.4)$$

Thus, the actual concentration achieved by the lens alone is

$$C_p = 1/(4f^2\sin 2\theta_p)$$
$$= \sin^2\phi_c / \sin^2\theta_p \qquad (6.5)$$

Thus, for practical lens systems in which $f$ is approximately >1.0, $C_p$ falls short of the limit of Eq. (6.2) by a factor of $1/4f^2n^2$. This is illustrated in Figure 6.2, in which the geometric concentration ratio relative to $1/2\sin\phi_c$ is plotted as a function of $f$.

The ideal limits for $n = 1$ and $n = 1.5$ are represented by horizontal lines in the figure, whereas the primary concentration falls off as $1/f^2$.

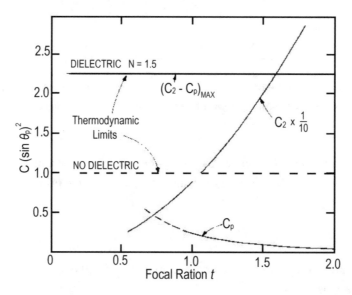

**FIGURE 6.2:** Achievable geometric concentration ratio as a function of the focal ratio of the primary lens.

If a DCPC-type secondary with entrance aperture diameter $L$ and exit aperture diameter $S$ is placed at the focal spot of the imaging primary lens, it can achieve an additional geometric concentration $C_2 = L^2/S^2$, which is given by Eq. (1.2b), with $\phi_c$ replacing $\theta_c$, or

$$C_2 = n^2/\sin^2 \phi_c$$
$$= 4n^2 f^2 \qquad\qquad (6.6)$$

The secondary concentration for $n = 1.5$ (renormalized downward by a factor of 100) is also shown in Figure 6.2. It can be seen that, in principle, the two-stage system can attain an overall geometric concentration equal to the ideal limit. In practice, certain compromises that somewhat reduce this need to be made, but, typically, secondary concentrations in the range 5–10× are readily achievable.

## 6.1.2  Secondary Designs

To illustrate the variety of second-stage designs that are possible, Figure 6.3 shows cross-sectional profiles for three cases. All profiles are designed for the same absorber diameter (i.e., cell size),

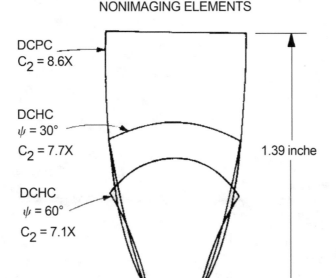

**FIGURE 6.3:** Representative solutions for nonimaging dielectric ($n = 1.5$) secondary profile shapes. All are designed for an $f/1.0$ primary ($\phi_c = 30°$) allowing $C_{2,\text{max}} = 9.0$.

here taken to be a circle of diameter $S = 0.25$ in (0.64 cm), and all are designed to be matched to a primary of focal ratio $f/1.0$ ($\phi_c = 30°$). The secondary is formed from a transparent dielectric with index $n = 1.5$ such that $C_{2,max} = 9×$. For the basic CPC design, in such a case, the totally internally reflecting condition is fulfilled only for

$$\sin\phi_c \leq (n - 2/n) \qquad (6.7)$$

or $\phi_c \leq 9.5°$, which is too small for most two-stage applications. Imposing the totally internally reflecting condition as a subsidiary constraint on the "maximum slope principle" yields a CPC solution of the $\phi_1 \geq \phi_2$ type (Rabl and Winston, 1976), with $\phi_2 = 76°$. $C_{max}$, or, in this case, $C_{2,max} = 8.6$ instead of 9.0. This is the DCPC solution shown in Figure 6.3. It has a height-to-aperture ratio of 1.9, which may be somewhat cumbersome to implement in a practical design and requires a substantial amount of dielectric material.

These problems can be alleviated for a small additional sacrifice in concentration by introducing an alternative design in which the front surface is curved so that it acts as a lens. The extra concentration provided by this lens changes the profile somewhat so that a more favorable aspect ratio results. In this case, in the limit of an aberration-free lens, the new profile would be a compound hyperbola, and so these solutions are referred to as DCHCs (Figure 6.4). In practice, for spherically

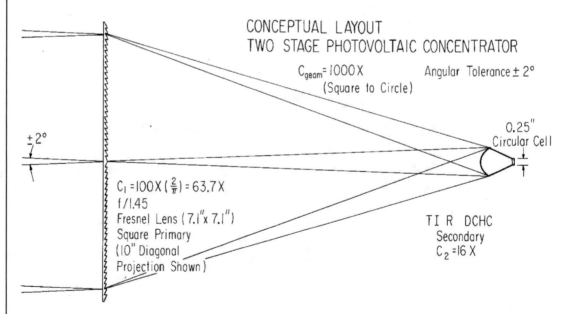

**FIGURE 6.4:** Schematic illustration of a possible design for a high-concentration (1000×) two-stage Fresnel lens DCHC concentrator.

curved front surfaces, the solutions found are not true hyperbolas. Clearly, an infinite variety of solutions are possible depending on the curvature of the surface actually employed. Two solutions shown are parameterized by $\psi$, the half-arc of the circle defining the entry surface. Cases for $\psi = 30°$ (i.e., the same as the convergence angle of the primary) and $\psi = 60°$ are illustrated (Ning et al., 1987).

## 6.2    SINGLE LARGE-SCALE PV CONCENTRATORS USED WITH A MULTICELL ARRAY

We have seen that at the present times, there are significant limitations to the concept of concentrating PV systems. High flux and nonuniform irradiance on the cell increase costs and impose the additional requirement that each cell has to have its own concentrator. Thus, virtually all concentrating PV systems developed up to now employ modular arrays of small concentrators combined with individual cells. This, in turn, requires that optics of the entire array be mutually aligned and tracked together to follow the sun.

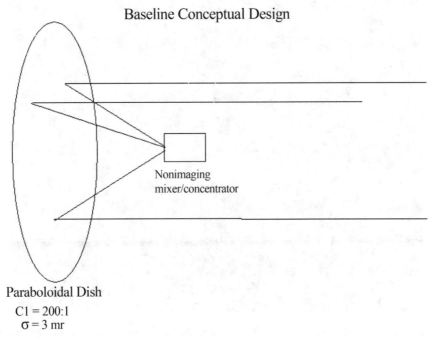

Baseline Conceptual Design

Nonimaging
mixer/concentrator

Paraboloidal Dish
$C1 = 200{:}1$
$\sigma = 3$ mr

**FIGURE 6.5:** Schematic illustration of a possible design for a scaled-up dish concentrator for a PV array. The secondary element is primarily for redistribution of the concentrated flux on the array, but nonimaging techniques to incorporate additional concentration in the mixer have been explored (O'Gallagher, Winston, and Gee, 2001).

There have been several attempts to circumvent these limitations. One new approach was suggested by Feuermann and Gordon (Solar Energy, 2001), who proposed using miniature dish concentrators with transparent dielectric "kaleidoscope" optical mixers to couple the flux to individual cells and alleviate the effects of nonuniformities. In another approach with some promise (O'Gallagher, Winston, and Gee, 2001), the entire concentrator can be scaled up and coupled to a larger array of cells (Figure 6.5). However, this is practical only if optical mixing can be employed to distribute the flux nearly uniformly over a multicell array. If this can be done while incorporating some further concentrations in the optical mixer utilizing simple nonimaging concepts, some further benefits can be realized (Figure 6.6).

In order to make feasible a practical system combining a single large-scale concentrator with an array of cells, the irradiance in the target plane needs to be quite uniform. Previous work investigating the possibility of using various secondary devices to accomplish this task, such as a cylindrical refractive optical mixer and a CPC-shaped square refractive mixer, suggested that a truncated

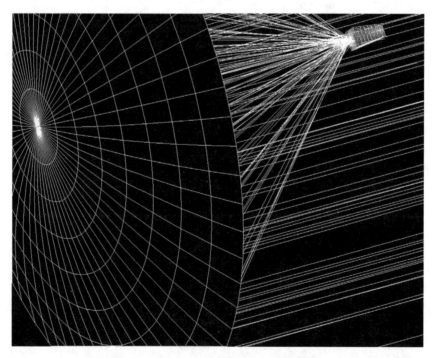

**FIGURE 6.6:** Conceptual design for a high-concentration scaled-up dish concentrator for a PV array (O'Gallagher, Winston, and Gee, 2001). Here, the primary optics is assumed to have an optical error defined by a $\sigma$ slope = 3 millirad, resulting in a primary concentration $C_1 = 800\times$. The secondary is a solid nonimaging dielectric pyramidal concentrator that both smoothes the concentrator flux and achieves a further concentration of $C_2 = 2.5\times$ such that the overall system concentration is $C_1 C_2 = 2000\times$.

square pyramidal totally internally reflecting optical mixer and concentrator do a good job of both optical mixing and concentrating. Although somewhat lower in concentrating power, a hollow reflective square pyramidal optical mixer and concentrator can also achieve these goals (O'Gallagher, Winston, Dunn, and Gee, 2002), such that an array of PV cells used with a single large reflective concentrating primary dish and secondary should be able to operate at high efficiency.

. . . .

CHAPTER 7

# Selected Demonstrations of Nonimaging Concentrator Performance

The optical response characteristics of CPCs interact with the direct and diffuse insolation components in ways that are unique. CPCs are not flat plate collectors. Neither, of course, do they behave like tracking troughs or dishes. Although the ideal optical response characteristics of CPCs are well understood formally and although experimental results for a wide variety of "one-off" prototype collector designs have been reported, there is a surprising dearth of information about the real-world optical performance of multimodule arrays of these collectors. To begin to remedy this situation, this monograph reports on the optical and/or thermal performance of three demonstration projects using CPC arrays that were comprehensively monitored for extended periods. These are as follows:

1.  a nonevacuated 3× CPC array installed on an elementary school (the Breadsprings Indian School) on the Navajo reservation near Gallup, New Mexico, in the late 1970s;
2.  an array of 1.3× evacuated CPCs instrumented and operated on the University of Chicago physics building during the late 1980s and throughout the 1990s; and
3.  an array of ~1× ICPCs used to drive a double-effect cooling system in a small office building in Sacramento, California, during the late 1990s and early 2000s.

## 7.1 THE BREADSPRINGS INDIAN SCHOOL NONEVACUATED CPC PROJECT

We have already mentioned, in Chapter 3, a 3× nonevacuated CPC designed for a vertical fin absorber. This collector was the prototype for a collector demonstration project undertaken by the University of Chicago in collaboration with the Bureau of Indian Affairs in the late 1970s. An array of these collectors was installed on the roof of the Breadsprings elementary school on the Navajo Indian Reservation about 25 miles south of Gallup, New Mexico.

The basic module design is shown in profile in Figure 7.1. The absorber in each trough is a 6.35-cm (2.5 in) vertical fin coupled to a copper tube. Both sides of the fin are utilized so that the 3× geometric concentration yields a 38-cm (15 in) aperture width for each trough and there are two such troughs, each 183 cm (6 ft) long, per module.

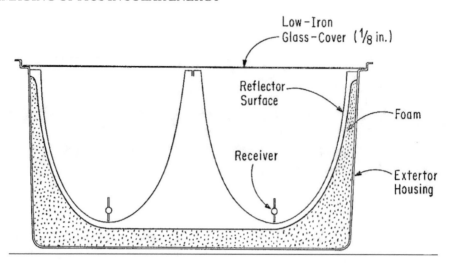

**FIGURE 7.1:** Cross-section profile of the 3× vertical fin collector module used in the 56-collector array at the Breadsprings school project.

The array was composed of seven rows of 8 collectors each, for a total of 56 individual collectors and a net collector area of 73 m² (785 sq ft).

The collector had an acceptance angle of 18° and was oriented with the collector normal tilted down relative to the vertical by latitude (37°) plus 16° = 53°. This allows the collector to "see" the sun through the 6 months of the year centered on the winter solstice. The measure acceptance properties of the collector and the collector orientation are illustrated in Figure 7.2, and a typical all-day performance near the winter solstice is shown in Figure 7.3.

## 7.2    THE UNIVERSITY OF CHICAGO EVACUATED XCPC ARRAY

Operating data for several years were acquired from a small array of commercial XCPCs installed in Chicago in the late 1980s. These data provided the first direct measurement (O'Gallagher, Winston, and Dallas, 2004) of the optical acceptance characteristics of a representative stationary CPC with a design acceptance half-angle of ±35° and a geometric concentration ratio of C = 1.33× (Figure 7.4). The observed incidence angle modifying functions in both the transverse and longitudinal planes are summarized and discussed subsequently. Of particular note is the fact that we found that the CPC collector "sees" about 92% of diffuse radiation rather than about 75%, which is predicted by models in which the diffuse component is assumed to be isotropic. The collector array functioned well for more than 18 years before it was removed after it became shaded by a new multistory building constructed adjacent and just to the west.

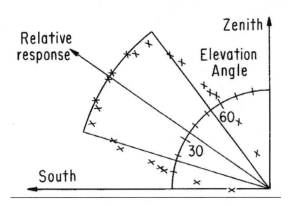

**BREADSPRINGS**
**3X CPC**
Angular Response
Shown for
fixed Winter
Collection Position

**FIGURE 7.2:** The 3× collector orientation (53° with respect to the vertical at the latitude of 37°) and measured acceptance angle (18°) are designed to collect solar energy for 6 months of the year around the winter solstice. If thermal energy was required all year, the collector would be repositioned for a summer orientation (21° with respect to the vertical).

These performance data provide direct measurements of several optical characteristics of CPCs that are critical to the accurate modeling of long-term performance projections for these collectors. These characteristics are (1) the transverse and longitudinal IAMs and (2) the "loss-of-diffuse" parameter.

The troughs of this early commercial CPC collector (based on the original Argonne National Laboratory design) have a geometric concentration ratio $C = 1.33\times$ (aperture width-to-absorber circumference), and the design acceptance angle (angular cutoff) is ±35°. The array of four panels, each with 12 horizontal (east–west aligned) troughs, was installed as recommended facing directly south and with the normal to the aperture plane tilted with respect to the zenith at the latitude angle for Chicago (42°). This optical design and its deployment are representative of those adopted for most stationary CPC designs so that the measured angular and diffuse acceptance properties can be used to characterize the performance of virtually all such stationary CPCs (O'Gallagher and Winston, 1983).

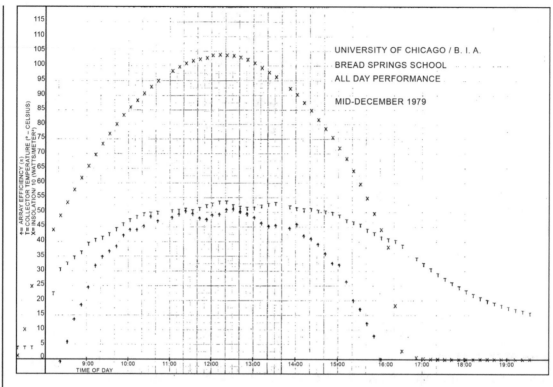

... All day performance curve near winter solstice showing available hemispherical insolation, average collector array temperature and array efficiency as functions of time. The fluctuations in temperature and efficiency near mid-day show the effect of load variations correlated with lunch time activities, e.g., dish washing, clean up, etc. The all day efficiency (8:30 to 16:30) is 39%.

**FIGURE 7.3:** The all-day performance curve for the full Breadsprings array near the winter solstice showing available hemispherical insolation, average collector array temperature, and array efficiency as functions of time. The fluctuations in temperature and efficiency near midday show the effects of load variations correlated with lunchtime activities. The all-day efficiency (8:30 to 16:30 h) is 39%.

The all-day performance, which helps separate the angular properties of the concentrator geometry from other effects, as measured on a very clear day near the summer solstice, is shown in Figure 7.5. At any given time, the sun's rays are incident on the collector aperture at an angle $\theta_I$ with respect to the aperture normal, which lies in the north–south (meridian) plane tilted at the latitude angle with respect to the zenith. It is conventional to characterize the angular response by resolving this incident direction further into the components $\theta_T$ and $\theta_L$, projected respectively onto the transverse and longitudinal planes containing the aperture normal. CPCs achieve stationary concentration in a trough geometry when tilted at the latitude angle and deployed with the longitudinal axis along the

**FIGURE 7.4:** Data from this four-panel array of commercial evacuated tube XCPCs should be very representative of most stationary CPCs. The design acceptance angle is ±35°, and the geometric concentration is 1.33×.

east–west direction so that the transverse angular acceptance is centered on the latitude angle. Thus, the sun at equinox will move along a path that lies in the center of the "wedge" of acceptance ($\theta_T = 0°$) throughout the day. On the other hand, at solstice, the sun will be at $\theta_T = ±23°$ (+23 in summer and −23 in winter) at noon and at larger angles throughout the rest of the day. The design acceptance of $\theta_c = ±35°$ is chosen to allow full collection for a minimum of 7 hours a day at solstice. The data in Figure 7.5 show a peak efficiency at noon near the expected value for $\eta_o$ with a gradual falloff on either side of noon and a relatively sharp cutoff around ±3½ hours (210 minutes) with respect to noon. Note that the turn-on in the morning is slightly rounder than the turn-off in the afternoon, although the sharp transition is still evident. This is due to a small amount of shading from the main wing of the building, which is to the east of the collector array. If both the longitudinal and transverse IAMs were equal to unity (and there was no shading) , the efficiency would rise sharply to its peak value as the sun entered the transverse acceptance before noon and then remain constant until 3½ hours after noon. Deviations from this idealized all-day response provide a direct measure of the IAMs.

The data in Figure 7.6 show the measured optical efficiency of this commercial CPC array as a function of the transverse incidence angle, $\theta_T$, near the summer solstice when the sun

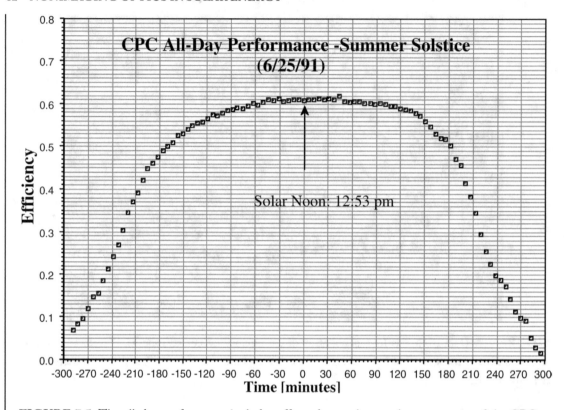

**FIGURE 7.5:** The all-day performance includes effects due to the angular acceptance of the CPC as well as the longitudinal and transverse IAMs.

moves through the angular cutoff at +35° twice a day. Once inside the design acceptance angle, the efficiency gradually approaches its peak value around noon as the transverse angle never gets much below $\theta_T = +23°$. These data clearly show the rounding of the ideal cutoff response due to mirror slope errors and partial throughput outside the acceptance angle due to truncation (Collares-Pereira, O'Gallagher, and Rabl, 1978).

One of the main advantages of CPC collectors relative to other concentrating collectors is their ability to collect a significant fraction of the diffuse radiation. If the diffuse component of insolation is isotropic, as is assumed in the simplest insolation models, it can be shown that the fraction of the diffuse that will be "seen" by a collector of geometric concentration $C$ is $1/C$ and the loss-of-diffuse factor will be $(1 - 1/C)$. From this, it follows that the factor $\Gamma$ in Eq. (3.4) is given by

$$\Gamma = [1 - f(1 - 1/C)] \tag{7.1}$$

where $f$ is the fraction of the total insolation due to diffuse radiation. Rewriting Eq. (3.4) and rearranging, we can express the dependence of the CPC optical efficiency on the diffuse fraction as

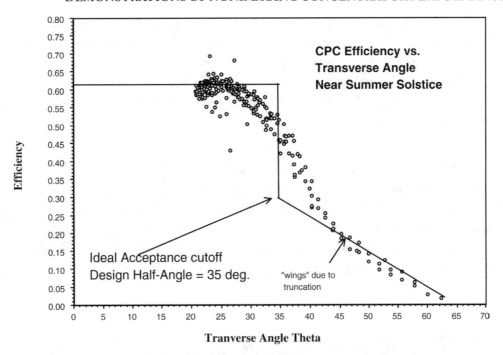

**FIGURE 7.6:** The ideal acceptance angle for a stationary CPC is compared with actual data near the solstice as the transverse incidence angle $\theta_T$ passes through the design cutoff toward its noontime value of just less than 23°. The rounding near 35° is expected due to random slope errors of the order 3°–5° and does not affect the long-term performance.

$$\eta_o(f) = \eta_{o,o} - \eta_{o,o} Lf \qquad (7.2)$$

where $\eta_{o,o}$ is $\eta_o/\Gamma$ and the loss-of-diffuse factor is $L = (C - 1)/C$.

It is well known that the diffuse component is not isotropic, being somewhat more intense at angles closer to the sun's direction, suggesting that a CPC might see more of the diffuse than predicted by simple models. A more sophisticated radiation model due to Perez (see Perez, Scott, and Stewart, 1983; Perez, Stewart, Arbogast, Seals, and Scott, 1986) gives some analytic support to this conjecture; however, until now, no experimental evidence for this behavior has been presented.

In an effort to explore this further, during 1991 and 1992, we conducted a set of measurements of the optical efficiency as a function of the diffuse fraction for a wide range of insolation conditions. In each case, the diffuse fraction was measured directly by shading a precision spectral pyroheliometer with a circular disk held so that it subtended a cone of half-angle 5.7° around the sun. The assumption of isotropic diffuse would lead to the prediction that with $C = 1.33$, the loss of diffuse should be $(0.33)/1.33 = 0.25$ and the CPC should collect about 75% of the diffuse insolation. With

$\eta_{o,o} \cong 0.60$, we should expect the slope of a fit to the optical efficiency as a function of diffuse fraction to be -0.15. Instead, the slope of the best-fit line is 0.053, or only about one-third of that predicted. To reiterate, these preliminary measurements indicated that, instead of "losing" 25% of the diffuse radiation, the CPC "loses" only about 8%, and we conclude that these collectors are "seeing" about 92% of the diffuse, substantially more than the 75% usually assumed in long-term performance projections for CPC collectors.

## 7.3   THE SACRAMENTO ICPC SOLAR COOLING PROJECT

Solar air conditioning represents a potentially huge market for solar energy in which the resource availability and load are well matched on both daily and seasonal bases. Analysis has shown that flat plate collectors cannot be economically viable for solar-driven absorption cooling under foreseeable conditions. Clearly, some kind of high-temperature collector is required even for single-effect absorption systems. For several decades, we have been promoting the use of evacuated CPCs, in particular, the

**FIGURE 7.7:** Closeup view of the fin absorbers and the end caps of some of the "easily manufacturable" ICPC tubes used in the Sacramento project.

ICPC, as the ideally suited solar collector for this application. In particular, there is a marked improvement in going to a double-effect system that should justify the further development of such systems.

A project to demonstrate the operational viability of the combination of the two concepts was initiated in 1995. This system was installed in 1998 and is still operating at the time of this writing (2008). The project and system are described more fully in several regular progress reports (see, for example, Duff, Winston, O'Gallagher, Bergquam, and Henkel, 2004; Duff, Daosukho, Winston, O'Gallagher, Bergquam, and Henkel, 2005; and references therein).

The ICPC design has been described in some detail in Chapter 3, and the collector design and cross-section were shown in Figure 3.7. A photo of a portion of the array under test is shown in Figure 7.7, and the thermal efficiency curve measured at the Sandia National Laboratory is shown in Figure 7.8.

A close-up of a portion of the ICPC array used in the Sacramento project is shown in Figure 7.9. The whole array was composed of 336 ICPC tubes of the design in Figure 3.7. They were arranged in three banks of 8 each for the 14-tube modules on the roof of an 8000-ft² one-story office building. Note that the CPC design had a very large acceptance angle effectively collecting from

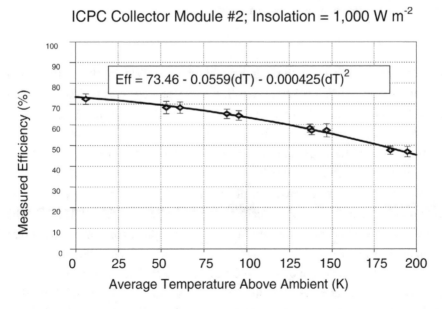

**FIGURE 7.8:** The thermal efficiency of a panel of nontracking (completely stationary) ICPC collectors of the design used in the Sacramento project as measured at the Sandia National Laboratory. Note that the efficiency remains above 50% at a temperature of 175 °C above ambient or above 200 °C for air conditioning applications.

**FIGURE 7.9:** Closeup view of a portion of the ICPC array used to drive a double-effect chiller cooling project in Sacramento, California, beginning in 1998.

the full sky. The array was aligned with the individual tubes parallel with the north–south direction. The total net area of the array is 1141 ft² (106 m²). Thermal storage was provided in a 3785-L (1000 gallons) tank. The whole array was used to drive a 20-ton, double-effect Li–Br/water absorption chiller. This chiller was commercially available but was modified to be hot water fired. Backup in case of cloudy days was provided by an auxiliary gas backup boiler (70 kW or 240,000 BTU/hour), which turned out not to be used very much.

The collector array was monitored instantaneously, and all-day efficiencies of each of the three banks were monitored separately. The collectors were typically operating at 150 °C, which corresponded to 110 °C above ambient on warmer days and 130 °C above ambient on cooler days, and the array performance was measured. A typical daylong performance curve is shown in Figure 7.10.

While operating in the range of 120–160 °C, daily collection efficiencies of nearly 50% and instantaneous collection efficiencies of about 60% were achieved in 1998 and 1999. Daily values of the Coefficient of Performance (C.O.P.) of 1.1 were achieved for the double-effect chiller. Two differently oriented fins (vertical or horizontal) gave essentially identical performances. The ICPC collector performance remained unchanged in nearly 2 years of operation. Eventual problems with the chiller pump resulted in reverting to a single-effect chiller (COPs of about 0.5–0.7) after 2002.

## Collection Efficiencies

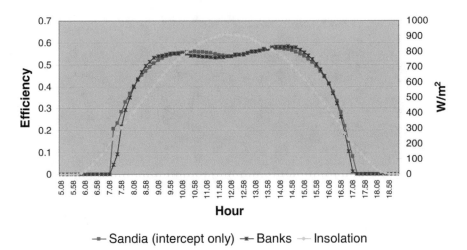

**FIGURE 7.10:** Typical clear day performance curve for the prototype ICPC collectors used to drive a double-effect chiller in a demonstration project in Sacramento, California.

The system easily met the full building load. The actual load varied between 12 and 16 tons as opposed to the rated capacity of 20 tons. This meant that the chiller was operating de-rated and under less-than-ideal conditions in that often the daylong operational performance was not optimized (collectors turned on late by hand on most days). Despite this, the system achieved typical daylong COPs of 0.9–1.1 even under a 12- to 16-ton load. It would be expected that the system would perform even better (COP ~1.2) under the projected full load.

In summary, this project has shown that

1. The high temperatures required for 2E chiller operation can be readily achieved with a nontracking evacuated concentrating collectors.
2. Reliable system (combined array–chiller) operation can be maintained over several years.
3. The overall system performance is four times that of a flat plate array operating a single-effect chiller. This corresponds to one-fourth the collector area for the same load!
4. The collector design is simple and readily manufacturable.
5. The collector has demonstrated the potential to be reliable over time scales of decades.
6. Solar air conditioning on a large scale for residential applications is a very realistic near-term possibility.

CHAPTER 8

# The Importance of Economic Factors in Effective Solar Concentrator Design

## 8.1 A RATIONAL MODEL FOR COST PERFORMANCE OPTIMIZATION

One very important aspect of designing any kind of solar system has to do with rational evaluation of cost and performance trade-offs (O'Gallagher and Winston, 2003). Often, when evaluating such concepts, a major emphasis is placed on developing a quantitative understanding of the technical performance, perhaps optimizing some standardized parameter related to conversion efficiency, such as optical quality and thermal absorption without regard for the related costs. However, although economic motivations are clearly present in considering these approaches, there is a tendency to be much less quantitative in attaining an understanding of the cost trade-offs involved in optimizing the system. Often, performance goals that are unattainable in practical economic systems are set and then used to design other parts of the system.

As has been previously noted (O'Gallagher, 1994), this practice of maximizing the efficiency with respect to some design parameter may not yield the most cost-effective configuration: That is, designs that allow the use of inexpensive materials and construction techniques may not (and probably will not) approach the performance of the most efficient systems one could build. Similarly, designs that attain the highest technically achievable efficiencies may not (and probably will not) be cost-effective. Despite the self-evident nature of these statements, one common approach has been simply to determine those parameter values required for maximum or near-maximum efficiency and select the corresponding designs as baseline or reference configurations. This practice unfortunately often results in selecting a development path, which leads away from the ultimate goal of minimizing the cost of delivered energy.

Here we consider a new methodology for the rational optimization of performance versus cost based on the constraint that, at the optimum, the relative incremental performance gains with respect to a particular performance parameter should balance the incremental costs associated with improvements in that parameter. We developed this methodology in the particular context of concentrating dish–thermal systems; however, the conclusions quantify some commonsensical relationships that should be applicable to many renewable energy systems. In this section, we present the elements of this model and discuss some of its implications.

## 8.1.1 The Model

The approach assumes that one has a system model that provides a quantitative measure of performance, say, efficiency $\eta$, as a function of some design parameter $u$. Furthermore, we assume that we know (or can reasonably guess) the relationship between this design parameter and the system cost $X$. The model shows that the rational economic optimum occurs for that value of the parameter $u$ for which the relative incremental efficiency gains are precisely balanced by the associated relative incremental cost increases: That is, optimum cost-effectiveness occurs at that value of $u$ for which the logarithmic derivatives of $\eta$ and $X$ are equal.

The fundamental objective of all solar system design is, of course, to maximize the energy delivered per dollar. The most sophisticated approach would be to deal with annual energy delivery and annualized costs. However, for a preliminary analysis, it should be sufficient to work with instantaneous efficiency and initial costs since these are directly related to the annualized quantities.

We begin by defining the quantity $R$, which is directly related to the energy per unit cost, as

$$R(u) = \frac{\eta(u)}{X(u)} \tag{8.1}$$

where $\eta$ is the instantaneous solar conversion efficiency under some fixed set of operating conditions (e.g., delivery temperature and insolation) and $X$ is the system cost per unit collection area. The quantity $u$ represents some design parameter on which both efficiency and cost depend. In our simple model, the most cost effective system will be that for which $R$ is a maximum.

Formally, we solve for the condition of maximum $R(u)$ by setting the derivative with respect to $u$ equal to zero as follows,

$$\begin{aligned}
\frac{dR}{du} &= \frac{1}{X}\left(\frac{d\eta}{du}\right) - \left(\frac{\eta}{X^2}\right)\left(\frac{dX}{du}\right) \\
&= \frac{\eta}{X}\left(\frac{1}{\eta}\frac{d\eta}{du} - \frac{1}{X}\frac{dX}{du}\right) \\
&= \frac{\eta}{X}\left(\frac{d(\ln\eta)}{du} - \frac{d(\ln X)}{du}\right) \\
&= 0
\end{aligned} \tag{8.2}$$

Thus, optimum cost-effectiveness occurs at that value of $u$ for which the logarithmic derivatives of $\eta$ and $X$ are equal: That is, when

$$\frac{d(\ln \eta)}{du} = \frac{d(\ln X)}{du} \tag{8.3}$$

Application of this model requires not only that both $\eta(u)$ and $X(u)$ are known but also that they can be represented by continuous and differentiable functions of $u$. In practice, these idealized conditions are unlikely to be met, particularly for the cost function. On the other hand, a great deal can be learned simply by representing the behavior by appropriate parametric models. Clearly, the optimum will depend strongly on the shape of both functions. It is important to note that $\eta$ is fundamentally bounded (it cannot be greater than unity!), whereas $X$ is not. It is quite possible that in

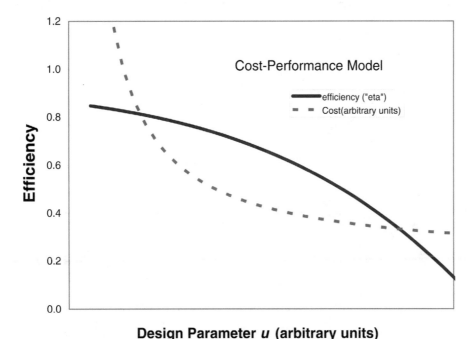

**FIGURE 8.1:** Schematic representation of a model for the rational optimization of performance and cost trade-offs. The cost and performance both increase with decreasing value of some design parameter u.

the very region in which $\eta(u)$ is approaching its limit, $X(u)$ will be increasing rapidly. If this is the case, maximizing $\eta(u)$ alone is clearly a misguided strategy.

The procedure is illustrated graphically in Figures 8.1–8.3 for a case in which both functions are monotonically decreasing with increasing $u$: That is, reducing the value of $u$ results in higher system efficiency, but this is achieved at a higher cost. It is the difference in the shape of these functions that determines the optimum.

In our illustration (Figure 8.1), the efficiency approaches a limiting value (here, 0.85) as $u$ goes to zero and falls off slowly as $u$ increases. In contrast, the cost $X$ is nearly constant at large $u$ but begins to increase relatively rapidly as it approaches zero. We have deliberately chosen these forms to emphasize the effects noted earlier; however, such a qualitative behavior is not at all unreasonable. The behavior of the ratio $R$ is shown in Figure 8.2, and both logarithmic derivatives are shown in Figure 8.3. Note that for such functions, the optimum is rather broad and occurs for values of $\eta$ that are significantly lower than its maximum value. To the left of the optimum (smaller value of $u$), marginal increases in cost more than offset the small incremental gains in efficiency, whereas to the right of the optimum (larger value of $u$), cost savings are too small to make up for the loss in performance.

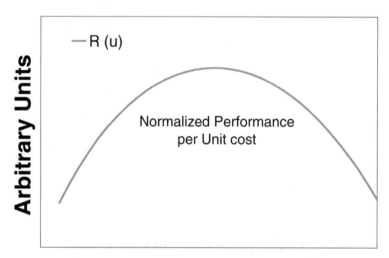

**Design Parameter *u* (arbitrary units)**

**FIGURE 8.2:** The maximum performance per unit cost [maximum $R(u)$] is quite broad and occurs for values of $u$ corresponding to significantly less than maximum performance.

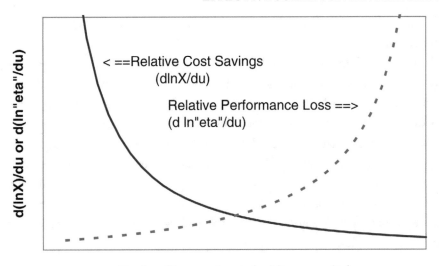

**Design Parameter *u* (arbitrary units)**

**FIGURE 8.3:** The value of *u* at which the logarithmic derivatives of the cost and performance functions cross (are equal) determines the cost performance optimum.

## 8.1.2    Conclusions

We have shown that the only rational optimization for a solar energy collection and conversion system is that for which the relative incremental efficiency gains are precisely balanced by the associated relative incremental cost increases—that is, when the logarithmic derivatives of the performance and cost functions are equal. Very generally, we conclude that the optimum for any system is likely to lie in a range of parameters in which both the costs and performance are varying significantly.

All cited data serve to emphasize the importance of understanding performance cost trade-offs in trying to choose among alternative paths toward the ultimate goal of low-cost solar energy.

In general, an approach such as that developed here can be particularly useful in defining both cost and performance goals. It emphasizes that accurate knowledge of the cost structure associated with all design trade-offs is essential both for choosing an optimized cost-effective design and for fully appreciating the relative merit of two-stage designs. Probably no single approach can provide the dramatic cost breakthrough still necessary for solar to compete favorably with conventional high-temperature thermal sources. However, perhaps a number of innovative technologies can be combined synergistically to deliver acceptable performance at greatly reduced cost.

## 8.2     THE ECONOMICS OF SOLAR COOLING SYSTEMS
### 8.2.1   Introduction
Solar cooling is one of the most naturally appealing applications for solar energy. The resource availability and the demand are roughly in phase, both on seasonal and diurnal time scales, and it is a "high-value application": That is, a British Thermal Unit (BTU) of cooling delivered by an electrically driven vapor-compression air conditioner costs the user substantially more than a BTU of heating supplied by burning natural gas. Why then, if the concept is so appealing, is solar cooling not already a fact of contemporary life? Why does the use of active solar cooling systems remain so limited? Clearly, there remain barriers to the practical implementation of solar cooling that are both technical and economic. In this section, we present a simple model (O'Gallagher and Winston, 1998) that helps quantify the economic factors and provides a basis for the comparison of different solar collector technologies.

### 8.2.2   A Simple Model
Our model assumes that the application is a solar-driven absorption cooling system displacing the electrical energy that would have been used to drive a conventional vapor-compression air conditioning system. It introduces a simple economic figure of merit that can be parameterized in terms of five dimensionless quantities, all of order unity. These five parameters are used to characterize (1) the climate, (2) the local cost of electrical energy, (3) the collector performance, (4) the performance of the proposed absorption cooling system, and (5) the performance of the existing electrical cooling system.

We introduce a basic figure of merit that is the approximate 10-year cumulative value ($V$) of the electrical energy displaced by the solar system. This 10-year value is a useful benchmark number since it can also be thought of as the maximum collector cost consistent with a simple payback of 10 years. This parameter can be calculated as follows

$$V = V_o \cdot EK\eta P \cdot (\text{C.O.P.})_{th} \qquad (8.4)$$

where

$$E = \frac{[\text{annual average insolation available to the collector}]}{170 \text{ W/m}^2}$$

$$K = \frac{[\text{cost of electricity}]}{\$ 0.10/\text{kW - hr}}$$

$$P = \frac{3.9}{(\text{C.O.P.})_{el}}$$

$\eta$ = annual average thermal collection efficiency of the collector array.

$(C.O.P.)_{th}$ and $(C.O.P.)_{el}$ are the coefficients of performance of the thermally driven absorption cooling chiller and the electrically driven vapor–compression system whose energy is being displaced, respectively.

(The coefficient of performance of a cooling system is the ratio of the quantity of heat removed from the cooled space to the energy that must be supplied to remove it.)

Note that $E$ depends on the location (i.e., climate) and the collector type and deployment geometry and $K$ depends on the location (i.e., local economic conditions) and particular application.

The three quantities, $E$, $K$, and $P$, in Eq. (1) have been defined as dimensionless ratios relative to typical representative values for the corresponding quantities. In particular, note that $E = 1.0$ corresponds to a yearly average insolation of 170 W/m², which in turn is what one would expect if one simply takes the local interplanetary solar constant of $I = 1370$ W/m², and takes a global average by multiplying by 1/4 [the ratio of the area of the earth's surface projected on a plane perpendicular to the earth–sun line $(\pi r_e^2)$ to the earth's global surface area $(4\pi r_e^2)$] and then again by a factor of 0.5 to account for weather effects. By multiplying by the number of seconds per year (about $\pi \times 10^7$ sec), one finds that this is equivalent to an annual average of insolation on a horizontal plane of 5.34 GJ/m²-year. The value of $E$ for a particular location and collector acceptance geometry then is a single parameter that adjusts up or down, usually by about a factor of 2, from this reference value. Similarly, the factors $K$ and $P$ are defined so that they are equal to unity for "typical" values of the cost of electricity and the C.O.P. for electrical air conditioning. Ten cents per kW-hr represents the midrange of electricity costs for summer peaking situations, and a coefficient of performance of 3.9 is standard practice for commercial vapor-compression systems. Note that there is nothing magical about these normalization values except that they have been chosen so that $V_o$ will be a representative "ballpark" quantity that characterizes the fundamental economics of solar cooling application. Note further that if other normalization quantities were chosen, it would modify $V_o$ but would not change the value of $V$ as determined by Eq. (8.4).

When the variables are normalized as defined, the quantity $V_o$ turns out to be equal to about \$380; that is,

$$V_o = \frac{10 \text{ years } 170(\text{W/m}^2)(8.76 \text{ kW - hr/W - year})\$0.10/(\text{kW - hr})}{3.9}$$

$$= 381.85(\$/\text{m}^2 - 10 \text{ years})$$

$$\cong \$380/\text{m}^2 \text{ - 10 years}$$

where, in view of all the other uncertainties, we have rounded $V_o$ to two significant figures. It is important to note that $V_o$ is a numerical constant *depending only on conversion factors* and is independent of assumptions other than the normalizations discussed earlier. These normalizations have been taken so that the dimensionless quantities in Eq. (8.4) ($E$, $K$, $\eta$, and $P$) have typical values ranging from ~0.5 to ~2.0 (of course, the efficiency $\eta$ is always less than 1.0) so that there is a relatively small range of possible values for our basic economic figure of merit $V$, typically within a factor of 4 of $V_o$. As a very rough rule of thumb, for even marginal economic viability, the solar collectors used to drive the absorption system should have a cost per square meter at most equal to $V$, and the yearly maintenance costs associated with the collectors must be a very small fraction of $V$ (i.e., $\lesssim 1\%$).

For example, although the thermal efficiency for a flat plate collector can be between 0.7 and 0.8 for heating domestic hot water, at temperatures around 100 °C, such as those required for driving a single-effect absorption chiller, its efficiency will generally be reduced to 0.1–0.3. The C.O.P. values for typical single-effect absorption chillers range from about 0.5 to 0.7. Thus, if for simplicity we take $E$, $K$, and $\eta$ all to be unity, we find that for a single-effect chiller driven by flat plates, $V$ turns out to be about \$20–\$80 at a typical mid-latitude location. As a practical matter, installed flat plates collectors cost substantially more per square meter than this (perhaps \$300–\$500/m²). Thus, it is clear that the low C.O.P. values attainable by single-effect chillers combined with the low thermal efficiency of flat plate collectors at even moderately high temperatures together make them unsuitable for this application and have contributed to the perception that active solar cooling is not economical. Significantly higher collector array operating efficiencies and higher chiller C.O.P. values will be required.

Recently, the manufacturers of air conditioning equipment have developed and are beginning to market commercial-scale double-effect absorption chillers. These units have C.O.P. values of about 1.2 (a factor of nearly 2 better than the best old single-effect units) but require heat to be delivered at about 160–190 °C, well beyond the capabilities of flat plate or ordinary evacuated tubular collectors. The ICPC (see Chapter 3) has the potential to increase the value of $V$ for solar absorption systems to a level closer to that required for economic viability. These collectors provide the only simple and most effective method for delivering solar thermal energy efficiently in the temperature range from 100 °C to about 300 °C without tracking. At 160 °C, they are expected to achieve efficiencies between 0.5 and 0.7, so that if, as above, we take $E$, $K$, and $\eta$ all to be unity, $V$ would be between \$230 and \$320, which is a factor of 5–10 better than that for flat plate collectors with single-effect systems and is at least in the right ballpark to have an appreciable chance for economic viability.

## 8.2.3   Discussion

This is admittedly an oversimplified model and probably has some unrealistic features; however, some definite conclusions are clear. Since even optimistic cost estimates for most collector types (even flat plates) lie in the range of $200–$400/m$^2$, it is clear that flat plate collectors cannot be economically viable under present-day conditions. Some kind of high-temperature collector is required even for single-effect absorption systems, and for any of these, the marked improvement in going to a double-effect system should justify the further development of such systems. The ICPC and the parabolic trough are quite comparable with each other in value delivered at most locations.

Finally, note that for the most encouraging of the cases analyzed, our value of $V_0 = \$380/m^2$ turns out not to be a bad representative ballpark figure for the cooling application. That this value is so low must be kept in mind in trying to develop an economically viable collector system (it is less than a penny a day per square foot!). This is important both in regard to the initial cost of the array and in that to the cost of maintenance. It is in this latter respect that the advantages of the nontracking, fully stationary ICPC relative to the parabolic trough could become paramount.

. . . .

CHAPTER 9

# Ultrahigh Concentrations

## 9.1 CONCENTRATION LIMITS FOR NONIMAGING TERMINAL CONCENTRATORS IN CENTRAL RECEIVER APPLICATIONS

### 9.1.1 Introduction

In this chapter, a preliminary analysis that carries out various performance trade-offs involved in the design of a two-stage central receiver plant that could achieve ultrahigh concentrations with a nonimaging CPC type secondary is presented. It should be noted that the study was carried out in the context of using a solar central receiver plant to generate hydrogen from the direct thermo-chemical splitting of water. However, the analysis is independent of that context and applies to any and all central receiver (or the so-called power–tower) configurations. This analysis shows that such a two-stage central receiver plant can achieve geometric concentrations approaching the maximum theoretical limits and provides a methodology for evaluating various geometrically dependent performance trade-offs. The concept of using a secondary (terminal) concentrator in central receiver systems has been around for some time (Rabl, 1976a; Athavaley, Lipps, and Vanthull, 1979) but has not been seriously investigated until relatively recently (Spirkl, Timinger, Ries, Kribus, and Muschaweck, 1998; Kribus, Huleihil, Timinger, and Ben-Mair, 2000; O'Gallagher, and Lewandowski, 2005).

The objective is to design a large-capacity (multimegawatt) plant, and it is clear that such a large scale with >1 MW power requirements can only be delivered by some kind of central receiver. We wish to consider the incorporation of a nonimaging secondary into such a configuration. In CPC designs, the geometry known to be optimal is a cylindrical, axially symmetric configuration, which, for a central receiver has a central tower surrounded by a circular area containing heliostats (Athavaley, et al., 1979) such as illustrated schematically in Figure 9.1a.

### 9.1.2 Limits to Central Receiver Concentration

The approach is based on simple geometry and the optical characteristics of CPCs. The goals are (1) to fill the field of view of the CPC as much as possible and (2) to identify those design factors

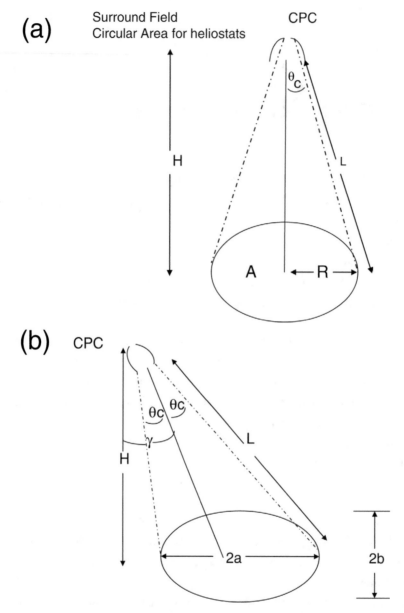

**FIGURE 9.1:** (a) The simplest geometry for a two-stage central receiver is a central tower (height $H$) surrounded by a circular heliostat field. The secondary is a simple CPC with acceptance angle $\theta_c$. The field radius is $R = H^*\tan\theta_c$. (b) Alternative geometry for a two-stage central receiver has a CPC whose optical axis is tilted at an angle $\gamma$ (toward the north in the northern hemisphere) to accommodate a heliostat field that has lower obliquity corrections when tracking the sun. The heliostat field intercepted by the CPC acceptance angle is elliptical in shape.

that either limit the achievable concentration or, on the other hand, allow the ideal limit on concentration to be approached as closely as possible.

The following general simplifying assumptions were made. No preconceived constraint on the basic configuration was imposed. The intercept of the view cone of the CPC was used to define the envelope of the area containing heliostats. Within this area, the fine-grained heliostat field losses were assumed to be represented by a field wide average. Only geometric losses were considered. Reflection losses were not taken into account. Blocking and shading losses were not calculated in detail. The angular distribution characterizing the incident solar radiation (i.e., sun size, slope, and specularity errors, etc.) was taken to be defined by a cone of a given half-angle (a "pill-box" distribution). The "maximum achievable concentration" was calculated by requiring that the optical intercept factor at the target for a particular configuration be 100%.

**9.1.2.1  Surround Field Geometry.** One can use simple geometry to determine the maximum solar image size for a particular configuration and for a particular half-angular subtense (including optical errors). From this, one can calculate the maximum achievable geometric concentration ratios $C_1$ and $C_{12}$, which can be attained without intercept losses, respectively for one- and two-stage configurations, and compare these with the ideal limiting concentration for the same angular subtense. These ratios, relative to the ideal limit, are plotted in Figure 9.2 for a circular surround field central receiver geometry as a function of the ratio of the tower height ($H$) to the diameter ($D$) of the circular field (Figure 9.1a). The concentration achievable by the field alone ($C_1$) maximizes at a value corresponding to one-fourth of the ideal limit at a tower height/diameter ratio of about 0.5. This is the same result that was found in Chapter 5 to apply to dishes, which have essentially the identical geometry to that shown in Figure 9.1a. If a CPC with a half-angle of acceptance just sufficient to "see" the entire field is placed in the target plane, it will achieve an additional secondary concentration of $C_2 = 1/\sin^2(\theta_c)$, where $\theta_c = \tan^{-1}(D/2H)$. The product of this with the field concentration ($C_{12} = C_1 C_2$), also plotted in Figure 9.2, will exceed that for the heliostat field alone and approach the ideal limit at large tower heights. Again, this is the same result as is well established for dishes (O'Gallagher and Winston, 1985). Note, however, that very tall towers are not needed, since even at tower height/diameter ratios near 1.0, the system concentration is over 80% of the ideal limit, more than three times the maximum achievable without a secondary. The CPC acceptance angle, $\theta_c$, for the corresponding two-stage system is also indicated as a function of the tower height/diameter ratio in Figure 9.2.

**9.1.2.2  Field Size.** We would like to estimate the scale of the required solar plant (i.e., the actual size of the heliostat field (and tower height, etc.). To do this, we combine the effect of all the loss mechanisms in a single parameter, $X$, such that the total power delivered to the tower is $XIA$, where $I$ is the direct

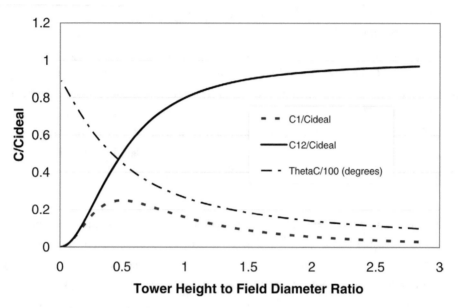

**FIGURE 9.2:** The maximum achievable geometric concentration relative to the thermodynamic limit for one- and two-stage central receiver systems in a surround field geometry as a function of the tower height-to-field diameter ratio. Also indicated is the corresponding secondary acceptance angle.

normal insolation and $A$ is the total area within the secondary field of view. This efficiency factor, $X$, will of course be a further function of solar incidence angles, location within the heliostat array, and material properties, among others. But for purposes of illustration, we will assume that these can all be approximated by some "average" over the array.

For a solar plant that would deliver a net total thermal power $P$, we will need a total area $A = P/XI$. If we rather arbitrarily take $X = 0.25$ and assume for simplicity that $I \sim 1000$ W/m², then, for a net solar thermal power of at least a megawatt, we will need a total ground area of about $A = 4000$ m². We use this to define a baseline configuration for comparison with alternative designs. From Figure 2, we are led to choose a two-stage central receiver design with a symmetric circular surround field geometry, $H/D$ near 1.0, and a CPC with $\theta_c = 25°$. For an area of 4000 m², this corresponds to a tower height of about 77 m, which is tall but not prohibitively so.

**9.1.2.3 Nonsurround Field Designs.** Although the baseline configuration defined earlier achieves 80% of the ideal limit for a given set of optical errors, it requires a relatively tall tower for the size of the heliostat field. One of the main purposes of this analysis is to explore the effect of deviating from this optimal axially symmetric configuration by tilting the CPC axis and aperture plane away from directly downward and pointing it at some "lookout angle" $\gamma$ off to the north. This alternative configuration is illustrated schematically in Figure 9.1b. The intercept of the view cone with the ground (assumed flat and horizontal) is an ellipse (a prototypical conic section). If one keeps the tower height constant, the area of the intercepted ellipse gets rapidly larger as $\gamma$ is increased.

For purposes of illustration, we rescale the geometry of the basic configuration as $\gamma$ is increased to keep the area constant, here, $A = 4000$ m². This might be regarded as a rather arbitrary procedure since so many other things (e.g., obliquity effects, blocking, and shading, among others) are also likely to change as $\gamma$ is varied. However, these effects can be incorporated later and this assumption keeps the scale of all the configurations comparable, so that the effect on other geometric quantities is directly manifest and easily interpreted.

In Figure 9.3, the effect on a number of geometric quantities of gradually tilting the CPC axis northward is plotted as a function of the lookout angle $\gamma$. The quantities plotted include the tower height, distance to the furthest heliostat, and the ellipse location. Also plotted is the maximum geometric concentration ratio achievable with the two-stage system. This has been calculated by first determining $C_{1,max}$, the maximum geometric concentration from the first stage, as determined directly from geometry and using the same approach as was used by Rabl (1976a). The maximum possible two-stage concentration ratio is then determined simply from

$$C_{12,max} = C_{1,max}{}^* C_{2,max} = C_{1,max}/\sin^2(\theta_c) \qquad (9.1)$$

This is the same approach used by O'Gallagher and Winston (1985) for point focus dishes and for the central receiver with a circular surround field plotted in Figure 2. In all these calculations, the geometric concentration ratio $C_{geom} = A_{col}/A_{abs}$ where $A_{col}$ is the effective collecting aperture of the system and $A_{abs}$ is the area of the target. Furthermore, in nonimaging optics, when we say that the concentration is "achievable," we mean that it applies to some directional distribution seen by $A_{col}$, such that the throughput to $A_{abs}$ is 100%—that is, it is without any geometric throughput losses. This means that $A_{abs}$ can be no smaller that the size of the sun's image produced by the heliostat most distant from the target. Finally, we note that $A_{col}$ is the effective area of the collecting aperture, including the effect of cosine factors. For configurations such as those schematically illustrated in Figure 1b, $A_{col}$ is not simply the gross area of the ellipse. Instead, it can be seen to be equal to the area of a circle *seen from the target* that would fill the same field of view as that of the ellipse. This is the circle defined by the intersection of the CPC acceptance cone with the plane perpendicular to the cone axis at the point at which that axis meets the ground. Substituting in all the appropriate geometric quantities, we find the scale independent result that

$$C_{12,max} = \cos^2(\gamma + \theta_c)/\cos^2(\gamma) \qquad (9.2)$$

which is plotted as a function of $\gamma$ in Figure 3. Note that, in the limit of a small $\gamma$, this quantity approaches the value for the circular surround field (as it must) and decreases steadily as one tilts the CPC plane away from directly downward and moves the heliostat field northward.

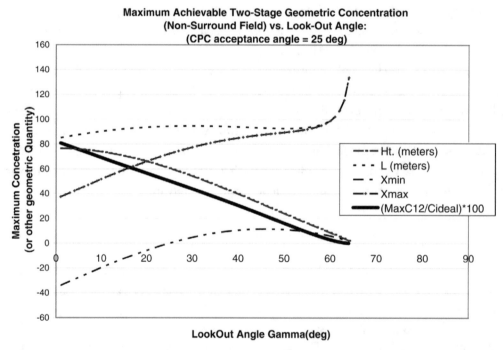

**FIGURE 9.3:** The maximum achievable geometric concentration as a percentage of the ideal limit is plotted as a function of lookout angle $\gamma$ for a baseline two-stage concept with a 25° CPC ($C_{max}$ = 5.60×). The variation of tower height and some other dimensions are also plotted. The area of the elliptical "footprint" intercepted by the CPC view angle is held constant at $A$ = 4000 m² as $\gamma$ is varied from 0° (the surround field limit) to 90° – $\theta_c$. The length $L$ is the distance from the target plane (the top of the tower) to the farthest point in the heliostat field, $X_{min}$ is the distance from the bottom of the tower to the southernmost boundary of the footprint ellipse, and $X_{max}$ is the distance from the bottom of the tower to the northernmost boundary of the ellipse.

**9.1.2.4  Obliquity (cosine) Effects.** The main reason for locating central receiver heliostat fields mostly to the north of the tower is to minimize cosine losses from the mirror area. In order to estimate the importance of these effects in this study, we have calculated the cosine correction in the meridian plane (the north–south plane containing the tower for a tower height of 60 m, a latitude of 45°, and for three noontime sun positions: winter solstice, equinox, and summer solstice). We find that the effect ranges from a maximum average cosine loss over the whole field of about 20% at winter solstice to about 5% at summer solstice. Note that the effect is not nearly as severe as it is for cases with a shorter tower relative to the field dimensions, such as most contemporary tower designs. It should also be noted, that although these effects are losses in cost-effectiveness, they do not result in a real loss in throughput or in concentration.

For a given set of optical tolerances, $C_{max}$ is determined by simple geometry and the distance $L$. This determines the relative size of the image of the sun at the target, which in turn determines the maximum achievable geometric concentration. Here it can be seen that the achievable concentration is greatest—about 80% of the ideal limit—for the surround field ($\gamma = 0$) and decreases steadily as the CPC is tilted to look out toward the north.

## 9.1.3  Summary and Conclusions

A preliminary analysis of various performance trade-offs involved in designing a two-stage central receiver plant secondary to achieve ultrahigh concentrations using a nonimaging CPC type has been carried out. The approach was based on simple geometry and the optical characteristics of CPCs.

We find that the highest possible concentrations can only be achieved with an axially symmetric circular field surrounding a central tower with a CPC looking vertically downward. In this configuration, the achievable concentration increases asymptotically toward the ideal limit as the CPC acceptance angle is reduced and the tower height increases. Although there is no well-defined optimum configuration, 80% of the ideal limit can be achieved in this configuration with a tower height-to-field diameter ratio of about 1.0. As one tilts the CPC view cone away from straight down, the intercepted area for heliostats becomes an ellipse of increasing area and eccentricity, and the maximum achievable geometric concentration decreases rapidly. (This result is independent of scale.) If the intercept area for heliostats is held constant, the corresponding tower height decreases gradually. Obliquity effects (cosine losses) are not prohibitively large in a circularly symmetric surround field geometry, mainly because the tower heights are comparable with the field dimensions. Effects of blocking and shading and ground cover constraints have not yet been calculated in detail but could be expected to be less in a surround field geometry than in a north field configuration. Furthermore, our approach guarantees that the calculated limit already includes the first-order correction for the effects of blocking and shading losses.

### 9.1.4  Recommendations

It must be emphasized that this is a preliminary study. The assumptions on which it is based are very idealized. For instance, the treatment of the incident angular distribution as a pillbox and the requirement that the secondary and target be sized so as to achieve 100% intercept are somewhat extreme. In practice, the optical errors are more likely to be better represented by some form of Gaussian distribution and the trade-off between concentration and intercept may yield relationships that differ in some details from those found here. However, based on the experience of applying similar arguments to the optimization of one- and two-stage parabolic dish systems (O'Gallagher and Winston, 1985, 1988b)[4], we are confident that the general conclusions reached here in this study regarding achievable concentration in particular configurations will serve as an excellent design guide for the actual behavior. With this in mind, we wish to offer the following recommendations:

1. The best designs for achieving high concentrations will be those with a circular surround field and a fairly tall tower.
2. The next best designs are those with $\gamma = \theta_c$ and as small a value of $\theta_c$ and corresponding tower height as are acceptable.

However, it should be noted that in the latter design, the achievable concentration would be reduced by a factor of approximately one-third.

## 9.2  SOME EXOTIC APPLICATIONS FOR ULTRAHIGH SOLAR FLUXES

From the earliest development of these nonimaging devices, when it became apparent that the thermodynamic limit on concentration could be approached, it was appealing to consider how one might practically develop very high levels of concentrated solar flux, in principle approaching even those found on the surface of the sun. However, pursuit of such objectives was deferred while the lower-concentration applications were developed.

Eventually, however, there occurred very rapid progress from the first ultrahigh flux measurements conducted on the roof of the high-energy physics building of the University of Chicago in 1988 to the experimental investigation of potential laser pumping and materials processing experiments carried out at the National Renewable Energy Laboratory High Flux Solar Furnace. The progression of demonstration experiments is summarized in Table 9.1. A comprehensive description of the techniques and history of these measurements is given by Jenkins, O'Gallagher, and Winston (1997).

Some of the potential applications for ultrahigh solar flux concentration include the following:

1. production of exotic materials (e.g., Fullerenes);
2. hydrogen production (direct water splitting);

TABLE 9.1 Experimental Measurements of Ultra-high Solar Flux Concentration Levels

| DATE | LOCATION | SECONDARY | MEASURED FLUX (SUNS) | TOTAL POWER |
|---|---|---|---|---|
| February 1988 | Chicago | Lens-oil filled Silver vessel ($n = 1.53$) | $56,000 \pm 5000$ | 44 W |
| March 1989 | Chicago | Solid sapphire DTIRC ($n = 1.76$) | $84,000 \pm 3500$ | 72 W |
| July–August 1990 | NREL (Golden, CO) | Water-cooled Reflecting silver CPC—air-filled ($n = 1.0$) | $22,000 \pm 1000$ | 3.5 kW |
| March 1994 | NREL (Golden, CO) | Fused silica (quartz) ($n = 1.46$) DTIRC with "extractor tip" | $50,000 \pm 2000$ | 900 W |

3. solar pumping of lasers;
4. high-temperature gas turbine solar receivers (Weizmann Institute for Science, Rehovath, Israel);
5. solar thermo-PV converters;
6. solar thermal propulsion in space; and
7. solar processing of materials in a lunar environment (or even in Mars).

The development of these techniques and applications is only one indication of the many fruitful benefits resulting from the concepts of nonimaging optics.

Many of the applications listed in the table for highly concentrated flux have been discussed elsewhere. However, the last two items that appear, application of solar concentrators in space and lunar environment, have not received much attention and may seem a bit bizarre. It is appropriate to consider here some of the unique advantages of nonimaging concentrators for such purposes.

## 9.2.1 Using Highly Concentrated Sunlight in Space

The techniques of nonimaging optics are particularly valuable in space or lunar environments in which the use of solar thermal energy has obvious advantages. Earlier preliminary studies have explored this concept for the production of cement from lunar regolith and for solar thermal propulsion in space. For example, extremely high temperatures, in the range of 1700–1900 °C, are necessary for the production of cement from lunar minerals. Such temperatures will in turn require very high levels of solar flux concentration. Energy budgets for the support of permanent manned operations on the lunar surface are expected to be limited. For high-temperature thermal (i.e., >300 °C) end uses, direct solar energy has obvious advantages over most other practical power sources. Conventional combustion processes are clearly impractical, and conversion of electricity (either solar or nuclear generated) to high-temperature heat represents a very wasteful use of high-quality energy. On the other hand, solar radiation is abundant and nondepletable. Most importantly, it is readily converted to heat with high efficiency, although at high temperatures this requires high concentration, as will be discussed subsequently. In the late 1970s, a small project was undertaken, with the support of the Jet Propulsion Laboratory, to investigate the potential of nonimaging designs for PV concentration in space. The project did establish that nonimaging design techniques could provide significant advantages and that the concepts were compatible with many other features thought to be important for deployment of concentrators in space (e.g., the use of large, lightweight reflecting membranes).

### 9.2.2   Applications in the Lunar Environment

In the lunar environment, solar energy is almost half again as intense as typical terrestrial levels (1350 W/m² versus about 900 W/m²) and nearly constant during the 2-week-long "lunar day." Of course, accommodation of the lunar night will preclude long-term, continuous solar-driven production and, lacking some kind of long-term thermal storage, will require some kind of two-phased monthly cycle. However, the energy requirements for processing lunar materials into cement are such that abundant long-term average production can easily be maintained. In particular, 1 kg of cement requires an input of about 1000 kcal. This means that even with a 50% duty cycle for the full lunar day and night, more than 100 tons of cement could be produced annually with a relatively modest solar furnace of roughly the scale we describe subsequently. Larger production would require either more or bigger furnaces.

### 9.2.3   Solar Thermal Propulsion in Space

Solar thermal propulsion systems in space will require very high temperatures to generate necessary levels of thrust by the direct solar heating and resulting expansion and expulsion of the propellant material. The generation of such temperatures, in the range 1400–2200 °C, will in turn require very high levels of solar flux concentration. In practice, to attain such levels, it may be useful and perhaps even necessary to incorporate some form of ideal or near ideal nonimaging concentrator. An analysis of the benefits associated with such a configuration deployed as a solar concentrator in the space shows that the thermal conversion efficiency at the temperatures required can be about three to five times that of the corresponding conventional design. Operational constraints on configurations that may be suitable for selected solar thermal propulsion applications are reviewed.

Although a variety of propellant materials and systems for direct solar propulsion in space are being considered, all will require very high temperatures, ranging from a low of about 1400 °C to more than 2300 °C. The efficiency for conversion of solar radiation to useful heat depends on the difference between the energy collected optically and that lost unavoidably through thermal processes driven by temperature differences between the hot absorber and its surrounding environment. Since these losses are small at low to moderate temperatures, it is comparatively easy to attain respectable efficiencies at temperatures below several hundreds of degrees Celsius. However, as one tries to operate at increasingly higher temperatures, radiation losses, which are proportional to differences in the fourth power of the absolute temperatures, increase very rapidly and can only be overcome by applying increasingly higher concentrations. The thermal efficiency of any concentrator collector system results from a balance between optical gain (typically between 75% and 85% due to reflection, absorption, and intercept effects) and thermal losses. Concentration can be used to reduce the latter. The consequences of the concentration limits for efficient solar thermal energy collection are illustrated in Figure 9.4. Here, the geometric concentration required to suppress

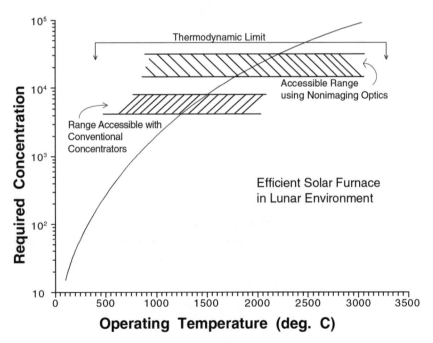

FIGURE 9.4: The geometric concentration required to achieve a respectable operating solar thermal conversion efficiency (~70%) in the near-earth space environment is shown as a function of the desired operating temperature. Also shown are the achievable concentration ranges for conventional and non-imaging concentrators.

thermal losses to about 5% of the solar gain so as to achieve a respectable operating solar thermal conversion efficiency (~70%) in the near-earth space environment is shown as a function of the desired operating temperature. Also indicated are the concentration levels attainable using either conventional concentrators or those that incorporate nonimaging designs. These levels are based on the optical quality thought to be achievable with reasonable cost for large state-of-the-art reflecting surfaces and correspond to an effective Gaussian slope error distribution with a standard deviation of 0.5–1.0 millirad. As can be seen, the temperature range for solar thermal propulsion requires concentrations that are unattainable (or only marginally attainable) with conventional designs and are only possible using some form of nonimaging concentrator.

.   .   .   .

# Bibliography and References

Athavaley, K., Lipps F.W., and Vanthull, L., An analysis of the terminal concentrator concept for solar central receiver systems, *Sol. Energy*, **22**, 493–504 (1979). doi:10.1016/0038-092X(79)90021-5

Bean, J.R., and Diver, R.B., The CPG 5-kW$_e$ DISH–Stirling Development Program, Paper no. 929181, Proceedings of the 27th IECEC, San Diego, CA (1992).

Cooke, D., Gleckman, P., Krebs, H., O'Gallagher, J.J., Sagie, D., and Winston, R., Sunlight brighter than the sun, *Nature* **346**, 802 (1990). doi:10.1038/346802a0

Collares-Pereira, M., O'Gallagher, J.J., and Rabl, A., Approximations to the CPC—a comment on recent papers by Canning and Shapiro, *Sol. Energy*, **21**, 245 (1978). doi:10.1016/0038-092X(78)90028-2

Collares-Pereira, M., O'Gallagher, J.J., Rabl, A., and Winston, R., A compound parabolic concentrator for a high temperature solar collector requiring only twelve tilt adjustments per year, Paper #1076, Proceedings of the International Solar Energy Congress, New Delhi, January 1978.

Dahl, J.K., Buechler, K.J., Weimer, A.W., Lewandowski, A., and Bingham, C., Solar–thermal dissociation of methane in a fluid-wall aerosol flow reactor, *Int. J. Hydrogen Energy*, **29**, 725–736 (2004).

Duff, W., Winston, R., and O'Gallagher, J.J., Cooling of commercial buildings with ICPC solar collectors, Proceedings of the 1995 ASME/JSME/JSES International Solar Energy Conference, Maui, HI, p. 1277, March 1995.

Duff, W.S., Winston, R., O'Gallagher, J.J., Henkel, T., Mushaweck, J., Christiansen, R., and Bergquam, J., Demonstration of a new ICPC design with a double-effect absorption chiller in an office building in Sacramento, California, In *Solar 1999*: the Proceedings of the 1999 ASES Annual Conference, Portland, ME (1999).

Duff, W., Daosukho, J., Winston, , R., O'Gallagher, J.J., Bergquam, J., and Henkel, T., Performance and reliability evaluation of the Sacramento demonstration novel ICPC solar collectors, in the Proceedings of the ISES 2005 Solar World Congress, Orlando, FL (2005).

Duff, W.S., Winston, R., O'Gallagher, J.J., Bergquam, J., and Henkel, T., Six year evaluation of the novel ICPC Sacramento demonstration, In *Solar 2004: A Solar Harvest, Growing Opportunities*: the Proceedings of the 2004 ASES Annual Conference, Portland, OR, July 2004.

Duffie, J.A., and Beckman, W.A., *Solar Engineering Thermal Processes*, Second Edition, John Wiley & Sons, New York, NY (1991).

Friedman, R.P., Gordon, J.M., and Ries, H., New high-flux two-stage optical designs for parabolic solar concentrators, *Sol. Energy*, **51**, 317–325 (1993). doi:10.1016/0038-092X(93)90144-D

Garrison, J.D., Optimization of a fixed solar thermal collector, *Sol. Energy*, **23**, 93 (1979a). doi:10.1016/0038-092X(79)90108-7

Garrison, J.D., Radiation collection by an optimized fixed collector, *Sol. Energy*, **23**, 103 (1979b). doi:10.1016/0038-092X(79)90109-9

Gleckman, P., O'Gallagher, J.J., and Winston, R., Concentration of sunlight to solar-surface levels using non-imaging optics, *Nature*, **339**, 198 (1989). doi:10.1038/339198a0

Gray, E.M., Possible cost advantage of a primary paraboloid/secondary hyperboloid tandem solar concentrator, *Sol. Energy*, **37**, 397 (1986). doi:10.1016/0038-092X(86)90029-0

Greenman, P., O'Gallagher, J.J., Winston, R., and Costogue, E., Reduction of intensity variations on photovoltaic arrays with compound parabolic concentrators, Proceedings of the International Congress of Solar Energy, Atlanta, 3, 1808 (1979).

Jaffe, L.D., Optimization of dish solar collectors with and without secondary concentrators, JPL Technical Report No. DOE/JPL-1060-57. U.S. Department of Energy (1982).

Jenkins, D., O'Gallagher, J.J., and Winston, R., Attaining and using extremely high intensities of solar energy with non-imaging concentrators, *Advances in Solar Energy*, Volume 11, 43–108 (Chapter 2), Karl W. Böer, Editor, American Solar Energy Society, Inc. (1997).

Jenkins, D., Winston, R., Bliss, J., O'Gallagher, J.J., Lewandowski, A., and Bingham, C., Solar concentration of 50,000 achieved with output power approaching 1 kW, *J. Sol. Energy Eng.*, **118**, 141–145 (1996). doi:10.1115/1.2870882

Kribus, A., Huleihil, M., Timinger, A., and Ben-Mair, R., Performance of a rectangular secondary concentrator with an asymmetric heliostat field, *Sol. Energy*, **69**, 139–151 (2000). doi:10.1016/S0038-092X(00)00046-3

Levi-Setti, R., Park, D.A., and Winston, R., The corneal cones of *Limulus* as optimized light concentrators, *Nature*, **253**, 115 (1975).

Luque, A., Non-imaging optics in photovoltaic concentration, *Phys. Technol.*, **17**, 118–124 (1986). doi:10.1088/0305-4624/17/3/I02

McIntire, W.R., Truncation of nonimaging cusp concentrators, *Sol. Energy*, **23** (4):351–355 (1979). doi:10.1016/0038-092X(79)90130-0

Mills, D.R., and Giutronich, J., Asymmetrical non-imaging cylindrical solar concentrators, *Sol. Energy*, **20**, 45 (1978). doi:10.1016/0038-092X(78)90140-8

Ning, X., Winston, R., and O'Gallagher, J.J., Dielectric totally internally reflecting concentrators, *Appl. Opt.*, **26**, 300 (1987a).

Ning, X., O'Gallagher, J.J., and Winston, R., The optics of two-stage photovoltaic concentrators with dielectric second stages, *Appl. Opt.*, **26**, 1207 (1987b).

O'Gallagher, J.J., Evaluation of performance and cost trade-offs in the optimization of two-stage solar dish electric systems, Proceedings of Solcom-I, The International Conference on the Comparative Assessments of Solar Power Technologies, Jerusalem, Israel, February 1994.

O'Gallagher, J.J., Comparative performance features of different nonimaging secondary concentrators, Proceedings of the SPIE Meeting, San Diego, CA (2001).

O'Gallagher, J.J., and Lewandowski, A., Achieving ultra-high solar concentration for the production of hydrogen in a central receiver plant using a nonimaging CPC type secondary, in Proceedings of the ISES 2005 Solar World Congress, Orlando, FL (2005).

O'Gallagher, J.J., Rabl, A., Winston, R., and McIntire, W. Absorption enhancement in solar collectors by multiple reflections, *Sol. Energy*, **24**(3):323–326 (1980). doi:10.1016/0038-092X(80)90490-9

O'Gallagher, J.J., Snail, K., Winston, R., Peek, S.C., and Garrison, J.D., A new evacuated CPC collector tube, *Sol. Energy*, **29**, 575 (1982). doi:10.1016/0038-092X(82)90067-6

O'Gallagher, J.J., and Winston, R., Development of compound parabolic concentrators for solar energy, *Int. J. Ambient Energy*, **4**, 171–186 (1983).

O'Gallagher, J.J., and Winston, R., Applications of maximally concentrating optics for solar energy collection, review chapter in *ENERGY SOURCES: Conservation and Renewables* (Hafemeister, Kelly, and Levi, Eds.), American Institute of Physics, AIP Conf. Proc. No. 135, pp. 448–471, New York (1985). doi:10.1063/1.35467

O'Gallagher, J.J., and Winston, R., Test of a "trumpet" secondary concentrator with a paraboloidal dish primary, *Sol. Energy*, **36**, 37–44 (1986). doi:10.1016/0038-092X(86)90058-7

O'Gallagher, J.J., and Winston, R., Performance model for two-stage optical concentrators for solar thermal applications, Proceedings of the 1987 Meeting of the American Solar Energy Society and the Solar Energy Society of Canada, Portland, OR, p. 89 (1987a).

O'Gallagher, J.J., and Winston, R., Performance and cost benefits associated with nonimaging secondary concentrators used in point-focus dish solar thermal applications, Solar Energy Research Institute Report, SERI/STR-253-3113-DE8801104 (1987b).

O'Gallagher, J.J., and Winston, R., Nonimaging concentrators (optics), review article in *Encyclopedia of Physical Science and Technology*, 9, 79 (Academic Press) (1987c).

O'Gallagher, J.J., and Winston, R., Evaluating the economic advantages of two-stage, tandem solar concentrators, *Sol. Energy*, **40**, 177 (1988a). doi:10.1016/0038-092X(88)90088-6

O'Gallagher, J.J., and Winston, R., Performance model for two-stage optical concentrators for solar thermal applications, *Sol. Energy*, **41**, 319–325 (1988b). doi:10.1016/0038-092X(88)90027-8

O'Gallagher, J.J., and Winston, R., Comparison of nonimaging secondary concentrators, In *Solar 1994*: the Proceedings of the 1994 American Solar Energy Society Annual Conference, pp. 179–183, San Jose, CA (1994).

O'Gallagher, J.J., and Winston, R., Development and test of a practical trumpet secondary concentrator for cavity receivers at high temperatures, Proceedings of the ISES 1997 Solar World Congress, Volume 2 (Solar Thermal), pp. 222–234, Taejon, Korea, August 1997.

O'Gallagher, J.J., and Winston, R., A simple economic model for solar cooling and the potential for the ICPC collector, in the Proceedings of the 1998 ASES Annual Conference, 401–406, Albuquerque, NM (1998).

O'Gallagher, J.J., and Winston, R., Practical design constraints for using secondary concentrators at high temperatures, in *Solar 1999*: the Proceedings of the 1999 ASES Annual Conference, Portland, ME (1999).

O'Gallagher, J.J., and Winston, R., A simple model for cost-performance optimization of concentrating dish systems, in *Solar 2003: America's Secure Energy*, the Proceedings of the 2003 ASES Annual Conference, Austin, TX, June 2003.

O'Gallagher, J.J., Winston, R., and Dallas, T., Long term performance of a first generation commercial external reflector evacuated tube CPC, in *Solar 2004: A Solar Harvest, Growing Opportunities*: the Proceedings of the 2004 ASES Annual Conference, Portland, OR (2004).

O'Gallagher, J.J., Winston, R., and Duff, W., Advanced solar cooling system with integrated CPC, published in *Harmony with Nature*, the Proceedings of the ISES Solar World Congress, Budapest, Hungary (1993).

O'Gallagher, J.J., Winston, R., and Gee, R., Nonimaging solar concentrator with near uniform irradiance for photovoltaic arrays, in *Nonimaging Optics: Maximum Efficiency Light Transfer VI* (R. Winston, Ed.), p. 60, Proceedings of the SPIE, Vol. 4446 (2001).

O'Gallagher, J.J., Winston, R., and Lewandowski, A., Optical properties of one and two-stage concentrator systems with non-paraboloidal and non-axisymmetric primaries, Proceedings of the 11th Annual ASME Solar Energy Conference, 195–200, San Diego, CA (1989).

O'Gallagher, J.J., Winston, R., and Lewandowski, A., Review of two-stage nonimaging concentrators for solar thermal power applications, Proceedings of the 1993 ISES Solar World Congress, Budapest, Hungary (1993).

O'Gallagher, J.J., Winston, R., and Welford, W., Axially symmetric nonimaging flux concentrators with the maximum theoretical concentration ratio, *J. Opt. Soc. Am.*, **4**(1):66–68 (1987).

O'Gallagher, J.J., Winston, R., Diver, R., and Lewandowski, A., Practical operation of a trumpet secondary concentrator with a cavity receiver at elevated temperatures, Proceedings of the 1997 ASES Annual Conference, Washington, D.C. (1997).

O'Gallagher, J.J., Winston, R., Diver, R., and Mahoney, A.R., Improved prospects and new concepts for secondary concentrators in solar thermal electric systems, Proceedings of the 1995 ASES Annual Conference, Minneapolis, MN (1995).

O'Gallagher, J.J., Winston, R., Diver, R., and Mahoney, A.R., Experimental demonstration of a trumpet secondary concentrator for the Cummins power generation (CPG) 7.5 kW$_e$ dish–Stirling system, Proceedings of the 1996 ASES Annual Conference, Asheville, NC (1996).

O'Gallagher, J.J., Winston, R., Dunn, L., and Gee, R., Reflective square pyramidal optical mixers for photovoltaic concentration, in the Proceedings of the 2002 ASES Annual Conference, Reno, NV (2002).

O'Gallagher, J.J., Winston, R., Mahoney, A.R., Dudley, V., Luick, D., and Gee, R., A new concentrating collector with evacuated all-glass absorber for solar cooling, in *Solar 2000*: the Proceedings of the 2000 ASES Annual Conference, Madison, WI (2000).

O'Gallagher, J.J., Winston, R., Mahoney, A.R., Dudley, V., Luick, D., and Gee, R., Thermal performance of an external reflector CPC with dewar-type evacuated glass absorber, in *Forum 2001*: the Proceedings of the 2001 ASES Annual Conference, Washington, D.C. (2001).

O'Gallagher, J.J., Winston, R., Zmola, C., Benedict, L., Sagie, D., and Lewandowski, A., Attainment of high flux-high power concentration using a CPC secondary and the long focal length SERI solar furnace, ASME Solar Energy Conference, Reno, NV, 337–343 (1991).

O'Gallagher, J.J., Winston, R., Suresh, D., and Brown, C.T., Design and test of an optimized secondary concentrator with potential cost benefits for solar energy conversion, *Energy*, **12**(3/4):217–226 (1987).

Ortabasi, U., Gray, E., and O'Gallagher, J.J., Deployment of a secondary concentrator to increase the intercept factor of a dish with large slope errors, Proceedings of the 5th Annual Parabolic Dish Review, Indian Wells, CA, DOE/JPL-1060-69, p. 170 (1984).

Perez, R., Scott, J.T., and Stewart, R., An anisotropic model for diffuse radiation incident on slopes of different orientations and possible applications to CPCs, Proceedings of the ASES Annual Conference, Minneapolis, MN, pp. 883–888 (1983).

Perez, R., Stewart, R., Arbogast, C., Seals, R., and Scott, J., An anisotropic hourly diffuse radiation model for sloping surfaces: description, performance validation, site dependency evaluation, *Sol. Energy*, **16**, 481–497 (1986). doi:10.1016/0038-092X(86)90013-7

Perkins, C., and Weimer, A.W. Likely near-term solar-thermal water spitting technologies, *Int. J. Hydrogen Energy*, **29**, 1587–1599 (2004).

Rabl, A., Comparison of solar concentrators, *Sol. Energy*, **18**, 93–111 (1976a). doi:10.1016/0038-092X(76)90043-8

Rabl, A., Optical and thermal properties of compound parabolic concentrators, *Sol. Energy*, **18**, 497 (1976b). doi:10.1016/0038-092X(76)90069-4

Rabl, A., Solar concentrators with maximal concentration for cylindrical absorbers, *Appl. Opt.*, **15**(7):1871–1873 (1976c).

Rabl, A., A note on the optics of glass tubes, *Sol. Energy*, **19**, 215 (1977). doi:10.1016/0038-092X(77)90062-7

Rabl, A., Goodman, N.B., and Winston, R., Practical design considerations from CPC solar collectors, *Sol. Energy*, **22**, 373–381 (1979). doi:10.1016/0038-092X(79)90192-0

Rabl, A. and Winston, R., Ideal concentrators for finite sources and restricted exit angles, *Appl. Opt.*, **15**, 2880–2883 (1976).

Rabl, A., O'Gallagher, J.J., and Winston, R., Design and test of non-evacuated solar collectors with compound parabolic concentrators, *Sol. Energy*, **25**, 335 (1980). doi:10.1016/0038-092X(80)90346-1

Ries, H., and Rabl, A., Edge-ray principle of nonimaging optics, *J. Opt. Soc. Am.*, **11**(10):2627–2632 (1994).

Ries , H., Spirkl, W., and Winston, R., Nontracking solar concentrators, *Sol. Energy*, 62, 113–121 (1998).

Ries, H., and Winston, R., Tailored edge-ray reflectors for illumination, *J. Opt. Soc. Am. A*, **11**, 1260–1264 (1994).

Snail, K.A., O'Gallagher, J.J., and Winston, R., A stationary evacuated collector with integrated concentrator, *Sol. Energy*, **33**, 441 (1984). doi:10.1016/0038-092X(84)90196-8

Spirkl, W., Timinger, A., Ries, H., Kribus, A., and Muschaweck, J., Non-axisymmetric reflectors concentration radiation from an asymmetric heliostat field onto a circular absorber, *Sol. Energy*, **63**, 23–30 (1998).

Suresh, D., O'Gallagher, J.J., and Winston, R., Heat transfer analysis for passively cooled "trumpet" secondary concentrators, *J. Sol. Energy Eng.*, **109**, 289–297 (1987).

Timinger, A., Spirkl, W., Kribus, A., and Ries, H., Optimized secondary concentrators for a partitioned central receiver system, *Sol. Energy*, **69**, 153–162 (2000). doi:10.1016/S0038-092X(00)00047-5

Welford, W., and Winston, R., *The Optics of Nonimaging Concentrators*, Academic Press, New York (1978).

Welford, W., and Winston, R., Design of nonimaging concentrators as second stages in tandem with image forming first-stage concentrators, *Appl. Opt.*, **19**(3):347–351 (1980).

Welford, W., and Winston, R., *High Collection Nonimaging Optics*, Academic Press, New York, NY (1989).

Winston, R., Light collection within the framework of geometrical optics, *J. Opt. Soc. Am.*, **60**(2):245–247 (1970).

Winston, R., Principles of solar concentrators of a novel design, *Sol. Energy*, **16**, 89 (1974).

Winston, R., Dielectric compound parabolic concentrators, *Appl. Opt.*, **15**(2):291 (1976).

Winston, R., Ideal flux concentrators with reflector gaps, *Appl. Opt.*, **17**(11):1668–1669 (1978).

Winston, R., Cavity enhancement by controlled directional scattering, *Appl. Opt.*, **19**, 195 (1980).

Winston, R., Nonimaging optics, *Sci. Am.*, **264**, 52–57 (1991).

Winston, R., Ed., *Nonimaging Optics: Maximum Efficiency Light Transfer III*, Volume 3139, Proceedings of SPIE—The International Society of Optical Engineering (1995a).

Winston, R., Ed., *Selected Papers on Nonimaging Optics*, SPIE Milestone Series 106 (1995b).

Winston, R., Cooke, D., Gleckman, P., Krebs, H., O'Gallagher, J.J., and Sagie, D., Brighter than the sun, *Nature*, **346**, 802 (1990).

Winston, R., Duff, W., O'Gallagher, J.J., Henkel, T., Christiansen, R., and J. Bergquam, R., Demonstration of a new type of ICPC in a double-effect absorption cooling system, in Proceedings of the 1999 ISES Solar World Congress, Jerusalem, Israel (1999).

Winston, R., Gleckman, P., Jenkins, D., and O'Gallagher, J.J., Ultra-high solar flux and applications to laser pumping, In *Solar Emerging, The Reality*, the Proceedings of the 1993 American Solar Energy Society Annual Conference, 225–227, Washington, D.C. (1993).

Winston, R., and Hinterberger, H., Principles of cylindrical concentrators for solar energy, *Sol. Energy*, **17**, 225–258 (1975). doi:10.1016/0038-092X(75)90007-9

Winston, R., and Ries, H., Nonimaging reflectors as functional of the desired irradiance, *J. Opt. Soc. Am. A*, **10**, 1902–1908 (1993).

Winston, R., and O'Gallagher, J.J., Performance characteristics of a two-stage 500× nonimaging concentrator designed for new high efficiency, high concentrator photovoltaic cells, Proceedings of the 1988 Annual Meeting, American Solar Energy Society, June 20–24, Cambridge, MA, p. 393 (1988).

Winston, R., O'Gallagher, J.J., and Gee, R., The gap problem in nonimaging solar concentrator design, in *Solar 2000*: the Proceedings of the 2000 ASES Annual Conference, Madison, WI (2000).

Winston, R., O'Gallagher, J.J., Mahoney, A.R., Dudley, V.E., and Hoffman, R., Initial performance measurements from a low concentration version of an integrated compound parabolic concentrator (ICPC), in the Proceedings of the 1998 ASES Annual Conference, 369–374, Albuquerque, NM (1998).

Winston, R., O'Gallagher, J.J., Muschaweck, J., Mahoney, A.R., and Dudley, V., Comparison of predicted and measured performance of an integrated compound parabolic concentrator

(ICPC), in the Proceedings of the 1999 ISES Solar World Congress, Jerusalem, Israel (1999a).

Winston, R., O'Gallagher, J.J., Muschaweck, J., Mahoney, A.R., and Dudley, V., Analysis of predicted and measured performance of an integrated compound parabolic concentrator (ICPC), in the Proceedings of the 1999 ASES Annual Conference, Portland, MD (1999b).

Winston, R., and Welford, W.T., Two-dimensional concentrators for inhomogeneous media, *J. Opt. Soc. Am.*, **68(3)**, 289–291 (1978).

Winston, R., and Welford, W.T., Geometrical vector flux and some new nonimaging concentrators, *J. Opt. Soc. Am.*, **69**, 532–536 (1979).

# Author Biography

**Joseph O'Gallagher** was born and raised in Chicago, IL. He received his undergraduate education at the Massachusetts Institute of Technology, where he earned a bachelor's degree in physics. He then returned to Chicago, where he received a master's degree and a doctorate degree (both also in physics) from the University of Chicago. He served on the faculty at the University of Maryland in the early 1970s and, for much of the time since then, was a senior scientist and administrator and lecturer at the Physics Department of the University of Chicago. His research work at Chicago was directed toward the development of practical and economical solar thermal and solar photovoltaic concentrators utilizing nonimaging optics. Dr. O'Gallagher has been an active member of the American Solar Energy Society for more than 30 years and recently completed serving 6 years on its Board of Directors. Before beginning work in solar energy, Dr. O'Gallagher spent more than 15 years working in experimental space physics. He and his wife Ellen have been married for 45 years and have two grown sons, four grandsons, and one (very special) granddaughter. Dr. O'Gallagher retired from the University of Chicago in 2005 and since then has been dividing his time among getting to know his grandchildren, consulting, volunteer work, writing this book, and teaching and lecturing. He has published more than 185 articles and technical reports in both experimental and theoretical areas and has lectured frequently on the topics of space science, solar energy, "peak-oil," and global climate change.